CULTIVATING KNOWLEDGE

CULTIVATING KNOWLEDGE
Genetic diversity, farmer experimentation and crop research

Walter de Boef
Kojo Amanor
Kate Wellard
with Anthony Bebbington

INTERMEDIATE TECHNOLOGY PUBLICATIONS
1993

Published by ITDG Publishing
103–105 Southampton Row, London WC1B 4HL, UK
www.itdgpublishing.org.uk

© CGN/ODI/WAU/Intermediate Technology Publications 1993

First published in 1993
Print on demand since 2003

ISBN 1 85339 204 9

A catalogue record for this book is available from the British Library

ITDG Publishing is the publishing arm of the Intermediate Technology
Development Group. Our mission is to build the skills and capacity of people in
developing countries through the dissemination of information in all forms,
enabling them to improve the quality of their lives and that of future generations.

Typeset by ODI and Inforum Typesetting, Rowlands Castle, Hants
Printed in Great Britain by Lightning Source, Milton Keynes

Contents

Introduction

Part 1.   Understanding Farmers' Knowledge

Part 2.   Developing Local Crops

Part 3.   Building Linkages

Part 4. Challenging Policy

Figures and Tables

Figures

Tables

Author biographies

Angela Cordeiro is a graduate in Agronomy. She works at AS-PTA, a Brazilian NGO working with plant genetic resources.

Martin Mbewe is a plant breeder by profession. He is a research co-ordinator of Thusano Lefatsheng, a local NGO for research, extension and marketing, Botswana.

Pat Mooney is the Executive Director of RAFI and has been working on agricultural biodiversity questions since the mid-1970s.

Melaku Worede is an agronomist (genetics and plant breeding) by profession, and he is retired director of the Plant Genetic Resources Centre/Ethiopia and current chairperson of the FAO Commission for Plant Genetic Resources.

Hailu Mekbib, agronomist, plant geneticist and breeder by profession, is the Head of Community-based Genetic Resources Activity Division of PGRC/Ethiopia.

David Millar is an agronomist and Agricultural Coordinator for the Catholic Archdiocese of Tamale, Ghana. He is presently doing research in Ghana for his PhD at Wageningen Agricultural University.

Jack Botchway is a Socio-Agronomist and an Extensionist by profession. He is currently the Managing Director of the Weija Irrigation Project in Ghana.

James Mascarenhas is an Indian specialist in the development of participatory approaches in rural development. He runs a organization called OUTREACH in Bangalore, South India, which addresses the issue of capacity building for small NGOs, and offers technical and managerial support and training.

Monica Atieno Opole is a Kenyan development worker. She conceptualized and implemented the KENGO Indigenous Vegetable Project. She is currently director of CIKSAP, a NGO which deals with local knowledge systems development.

Louk Box is a sociologist, who worked for the Wageningen Agricultural University in the Dominican Republican, studying the role of social sciences in agricultural research. He is presently director of the European Centre for Development Policy Management.

Dirk van Dusseldorp is Professor of planning and policy development at the Wageningen Agricultural University. He studies interdisciplinary cooperation in agricultural research.

Marianne van Dorp is a nutritionist and worked in the Technology Development for Farmers Project in Malang, Indonesia. Currently she is attached to Wageningen Agricultural University.

Ton Rulkens is an agronomist and worked in the Technology Department for Farmers Project in Malang, Indonesia, where he was responsible for the genebank. He is presently working at the Mondlane University in Mozambique.

Trygve Berg is trained as a plant breeder, but is currently working on development of agricultural production systems with particular interest in the management of genetic resources.

Mario Tapia is an agronomist, and a specialist on Andean farming systems and crops. He works as a consultant with different NGOs in Andean agriculture.

Alcides Rosas is a farmer in Peru. He has become a technician involved in projects on Andean crops and agriculture.

Gordon Prain is currently co-ordinator of the User's Perspective within Agricultural Research and Development (UPWARD), a Dutch-funded network supporting participative agricultural research in Asia. By profession, he is a social anthropologist with the International Potato Centre which sponsors the network.

Dan Taylor is a development agriculturalist by profession. He is director of the Centre for Low Input Agricultural Research and Development in Natal, South Africa.

Catherine Longley is an associate of the Department of Anthropology, University College London. She is currently researching local knowledge systems among small-scale rice cultivators in Sierra Leone.

Paul Richards is a geographer and anthropologist. He has done research on local knowledge and farmer experimentation in West Africa, especially Sierra Leone. He is presently Professor in Agrarian Change and Technology at the Wageningen Agricultural University.

Andrew Mushita is a social scientist and agronomist, and is manager of the indigenous seeds project with ENDA-Zimbabwe.

Saskia van Oosterhout is a plant biologist and is presently research associate at the Agricultural Research Centre in Zimbabwe.

Jaap Hardon is a plant breeder by profession. He is director of the Netherlands Centre of Genetic Resources and is involved in the international debate on plant genetic resources. He is currently programme-leader of the Community Biodiversity Development and Conservation Programme.

Walter de Boef is plant breeder by training, with a strong emphasis on plant genetic resources and links to social sciences. He is currently international coordinator of the Community Biodiversity Development and Conservation Programme based at the Centre for Genetic Resources, the Netherlands.

Anthony Bebbington is a geographer and Research Fellow at the Overseas Development Institute where he coordinates the Agricultural Research and Extension Network.

Kate Wellard was formerly a Research Fellow with the Overseas Development Institute. She is now Lecturer in the Department of Rural Development, Bunda College of Agriculture, University of Malawi.

Kojo Amanor is an anthropologist working on farming systems, local knowledge, and environmental issues at the Institute of African Studies, University of Ghana, Legon.

Preface

To understand what farmers know and do can provide crop researchers with better insights into agriculture and the agro-ecosystem. This book is an attempt to further this understanding. It is an early part of a new programme of community-level crop development and biodiversity conservation. The book presents a number of case studies from Africa, Latin America and Asia, which examine the significance of local knowledge, documenting new approaches and methodologies which have been developed for building linkages between farmers and researchers, and examining policy issues that stem from a concern for the negative repercussions of the expansion of agribusiness on the interests of small-scale farmers. The contributors come from a variety of backgrounds in the social and natural sciences and include NGO development workers and lobbyists.

Local knowledge and agricultural research

Earlier versions of the different papers in this book were first discussed at the seminar on Local Knowledge and Agricultural Research in Nyanga, Zimbabwe, 28 September – 2 October 1993. This was organized by the Wageningen Agricultural University (WAU) – Department of Sociology of Rural Development, ENDA – Zimbabwe, CPRO-DLO Centre for Genetic Resources the Netherlands (CGN) and Genetic Resources Action International (GRAIN). The seminar marked the conclusion of a research project of WAU Department of Sociology of Rural Development funded by the Netherlands Directorate General for International Cooperation (DGIS), which studied the potential contributions of social scientists to agricultural research and the role of farmers' knowledge systems in agricultural development in the Dominican Republic and the Philippines. The project showed the contradictions between the informal research carried out by farmers themselves and formal research in agricultural research institutes, and analysed appropriate strategies for adaptive crop research.

The main objectives of the seminar were to gain insight into the dynamics of local knowledge and the interaction between this knowledge and formal agricultural research systems, to document approaches and methods which have been devised for improving the linkages between the two systems, and to examine organizational and institutional frameworks and reforms needed for improving the dialogue between farmers and researchers. Some sixty people came together from a wide range of sectors, including non-governmental organizations active in the management and use of plant genetic resources, in sustainable agricultural development and in advocacy, managers of genebanks, plant breeders, and social scientists. The participants were mainly from Africa, with some people from Latin America, Asia, Europe and North America also contributing. One of the outputs of the seminar is the publication of the present book *Cultivating Knowledge* based on a selection of the papers presented at the seminar. The editing of the book was a joint project of CPRO-DLO Centre for Genetic Resources the Netherlands (CGN), WAU and the Overseas Development Institute (ODI).

The Community Biodiversity Development and Conservation programme

The seminar in Zimbabwe also served as a starting point for a research programme co-ordinated by CGN. Together with a number of non-governmental organizations, other institutes and genebanks, CGN studies and develops strategies for local crop development in Africa, validating local knowledge and promoting the use of local germ-plasm in Africa, Asia and Latin America. The Community Biodiversity Development and Conservation (CBDC) programme emphasizes the enhancement of farmers' capacity to conserve and develop local plant genetic resources. Major donors of the programme are DGIS, the International Development Research Centre (Canada) and the Swedish International Development Authority (SIDA). SIDA supported the seminar and publication of the present book as a starting point of the CBDC-Programme. The Swedish Agency for Research Cooperation with Developing Countries (SAREC) DGIS and the International Board on Plant Genetic Resources (IBPGR) also supported the organization of the seminar.

We would also like to acknowledge the freedom that Dirk van Dusseldorp and Louk Box of Wageningen Agricultural University gave us in deciding on the final structure of the book. At ODI, Peter Gee responded flexibly to our requests and was a great help in seeing the text through to its final form. Sandra Cox and Alison Saxby dealt very patiently with more editorial changes than they would care to remember.

The scope of the book

The papers prepared for the seminar spanned a range of approaches for thinking about, and working, in local crop development. They represented three broad perspectives on the relationship between farmers and researchers in local crop development. The first of these focuses on the socio-cultural environment in which local crop development occurs: the ideas and social structures that influence the pattern and process of farmer innovation. The second is more biological: its concern for farmer experimentation and crop management is motivated primarily by an interest in how the biological and genetic rationales underlying farmer resource use can contribute to improved processes of crop development. The third perspective examines policy and political economic frameworks in agricultural research and development, which all too often operate to the disadvantage of the small farmer.

The introductory chapter of the book analyses different approaches to local knowledge and crop development which have developed in the social and biological sciences and the rationale for their development. The first section brings together papers that share a dominantly socio-cultural and socio-economic focus on local crop development. They focus their attention on the social-economic and cultural context within which local knowledge is developed, managed and disseminated. Section two contains papers that focus on plant genetic resources and are interested in local knowledge in the context of developing crops and managing genetic resources. The third section is concerned with methods and mechanisms for building linkages between these different approaches in research and in action, uniting the social and the biological. The final section is ultimately concerned with the need to change the wider policy environment in such a way as to facilitate the incorporation of local concerns into crop development strategies, and to challenge the dominance of agricultural research by certain interest groups.

Walter de Boef
Kojo Amanor
Kate Wellard
Anthony Bebbington

London and Wageningen
August 1993

Abbreviations

AS-PTA	Consultants in Alternative Agricultural Projects
AVRDC	Asian Vegetable Research and Development Centre
CGIAR	Consultative Group on International Agricultural Research
CGN	Centre for Genetic Resources, Netherlands
CIAT	International Center for Tropical Agriculture
CIMMYT	International Maize and Wheat Improvement Center
CIP	International Potato Center
UPWARD	Users' Perspective within Agriculture and Rural Development
CLADES	Latin American Consortium for Agroecology and Development
CLIARD	Centre for Low Input Agricultural Research and Documentation
CPRO	Centre for Plant Breeding and Reproduction Research
DGIS	Netherlands Directorate-General for International Cooperation
DLO	Netherlands Agricultural Research Department
EMBRAPA	Brazilian Company for Agricultural Research
ENDA	Environment and Development Activities in the Third World
FAO	Food and Agricultural Organization of the United Nations
FSR	Farming systems research
GRAIN	Genetic Resources Action International
HEIA	High external input agriculture
IARC	International Agricultural Research Centre
IBPGR	International Board for Plant Genetic Resources
ICRISAT	International Crop Research Institute for the Semi-Arid Tropics
IDRC	International Development Research Centre
ILEIA	Information Centre for Low External Input and Sustainable Agriculture
IRRI	International Rice Research Institute
KENGO	Kenya Energy Non-Governmental Organization
LEISA	Low external input and sustainable agriculture
MARIF	Malang Research Institute for Food Crops
NGO	Non-governmental organization
ODI	Overseas Development Institute
PGRC/E	Plant Genetic Resource Centre /Ethiopia
PIDOW	Participative and Integrated Development of Watersheds Project
PRA	Participatory rural appraisal
PRATEC	Andean Project for Peasant Technology

PTD	Participatory technology development
RAFI	Rural Advancement Foundation International
RRA	Rapid rural appraisal
SAREC	Swedish Agency for Research Cooperation with Developing Countries
SIDA	Swedish International Development Agency
ToT	Transfer of technology
UNCED	United Nations Conference on Environment and Development
WAU	Wageningen Agricultural University

Introduction

Kojo Amanor, Kate Wellard, Walter de Boef and Anthony Bebbington

For thousands of years farmers have been adapting crops to diverse environments and experimenting with and developing new varieties. The interaction between people, the environment and their food crops has provided the world with a wide range of crops and a remarkable diversity of varieties within single crops. This diversity encompasses both the varieties farmers have selected, or landraces, and their wild and weed relatives. These interactions have also resulted in a human capacity to further develop crops through a process of continuous adaptation and experimentation.

Agricultural modernization, commercialization, intensification of production, and destruction of habitats are promoting genetic erosion, and threatening both this diversity of local crops and the processes which sustain it (Frankel, 1970; Harlan, 1984; Kloppenburg and Kleiman, 1987). This also results in a loss of farmers' knowledge of crops and of their capacity to maintain and develop diversity (Warren, 1991; NRC, 1992). Institutionalized crop breeding relies to a great extent on landraces originating in the major centres of diversity in the South. The genetic material available for this modern crop breeding is therefore being diminished. While genetic erosion threatens the world's base of food plants, the erosion of knowledge threatens the human capacity to maintain and further cultivate this diversity.

This collection of papers examines the threat to global agricultural diversity and the implications for agro-ecosystems. It addresses the need to develop appropriate research and development strategies which build upon both the capacities of farmers to experiment with crops, and the knowledge they have acquired of diversity. Farmers' experiences with diversity provide an important framework for the development of conservation strategies. An appreciation of these experiences can complement and add new dimensions to current conservation and crop improvement efforts, which frequently fail to realize the significance of interactions between farmers and environments in the development of biodiversity and emergence of local varieties. Farmer experiments with crops have been important in promoting diversity and the conservation of species and varieties. A challenge facing the agricultural sciences is to develop methodologies and institutional forms which will enable farmers to build upon their skills in adapting crops to the environment.

Landraces have made vital contributions to crop science, but the role farmers have played in their development has been largely unacknowledged. Apart from the need to understand and revalidate farmers' knowledge of crop development, this raises institutional and policy issues concerned with the value and appropriation of people's knowledge, the commercialization of plant breeding, and the marginalization of small-scale farming communities.

Farmers and diversity

In recent years there has been growing realization that human activities contribute considerably to genetic diversity. Conservation strategies can benefit from building linkages with the crop development strategies of rural people (Altieri and Merrick, 1987; Brush, 1991; NRC, 1992; Woods and Lenné, 1993 Oldfield and Alcorn, 1987; Vaughan and Chang, 1993). Farmers have created and managed environments where plants could evolve under selective pressure. These environments differed from those occurring in those 'wild' environments only marginally disturbed by people. They have adapted their farming practices, crops and varieties to different environments, thus creating a diversity of agro-ecosystems, crops and varieties (Richards, 1985). Such ecological and genetic diversity provides security for the farmer against pests, disease and unexpected climatic conditions (Lipton, 1968; Clawson, 1985). Diversity contributes to optimal production in highly variable and often marginal environments. Farming communities have historically adapted the potential diversity of crops and varieties to distinct and changing environmental conditions.

In contrast, modern agriculture is largely concerned with standardization and attempts to homogenize the environment to achieve optimal production conditions. It adapts the environment to homogeneous crops (monocultures, and hybrid or single line modern varieties), through technologies based on external inputs such as irrigation, fertilizers and pesticides. The aims are to optimize technological inputs and monocultural yields rather than local and natural resources, and to make environments uniform so that they respond to standardized technology. Modern varieties and hybrids seeds have become a driving force in the process of agricultural modernization and have made an enormous impact on agriculture in many regions of the world (Lipton with Longhurst 1989), threatening local diversity. The development of more environmentally sustaining technologies requires the transformation of these strategies. Greater emphasis needs to be placed upon preserving the diversity of varieties, crop species, agro-ecosystems, regions and societies (Keystone Center, 1991; NRC 1992; WRI et al., 1992; IPGRI, 1993).

Landraces, conservation and crop development

Landraces are the outcome of a continuous and dynamic development process. They are not stable products which have existed for time immemorial or which have remained static after coming into being. Friis-Hansen (1993) depicts landraces as the products of selective human interference in a dynamic process of adaptation to local agro-ecological production conditions, local sub-optimal production conditions, and to the specific production preferences of different socio-economic, gender and ethnic groupings within farming communities.

Farmers maintain, recombine and select landraces as intra-crop diversity. This process is influenced by the availability of an assortment of germ-plasm within a locality and the reproduction system of the crop. The techniques and efficiency in generating and maintaining local diversity within landraces differ considerably between self-pollinating (rice, wheat, bean), cross-pollinating (maize) and vegetatively propagated crops (yam, sweet potato, cassava and potato).

Landraces of self-pollinating crops are outbreeding crops, which means that certain characteristics are easy to maintain when the landraces have become homozygous through a process of mass selection. Recombination occurs (Richards, 1985), but on a less frequent basis than in cross-pollinating crops. Mixtures of different lines of crops like rice and beans are frequently cultivated and continuously improved (Conklin, 1957; Ferguson and Sprecher, 1989; Sperling et al., 1993). Farmers have an extensive knowledge of different varieties or varietal mixtures of rice, beans and other crops (Maurya et al., 1988; Voss, 1992).

Landraces of cross-pollinating crops are more difficult to maintain, because of the high risk of contamination with pollen of other genotypes. Friis-Hansen (1987) reports that in Tanzania farmers have devised maize breeding strategies based on the isolation of landraces in time and space. In contrast, in Mexico and Guatemala, farmers grow a wide range of maize varieties and have developed strategies in which maize varieties are cultivated in close proximity to weedy relatives, stimulating hybridization and introgression between maize and its relatives, through sophisticated recombination and seed selection procedures (Wilkes, 1977; Louette, 1991; 1992).

Fields of vegetatively propagated crops, such as potatoes, cassava, yams and sweet potatoes are dynamic evolutionary systems. Diversity in Andean potato is a product of conscious and active selection by farmers (Brush et al., 1981). Local diversity of cassava can be achieved by recombination of genotypes through strategies based on nurturing volunteer cassava seedlings (Boster, 1984; 1985).

Farmers have developed strategies to promote diversity for all the various forms of crop reproduction. These are based on specific processes of maintenance, recombination and selection. However, in many parts of the world, these strategies are now under threat and are disappearing at an alarming rate. If conservation strategies are to be developed further, it is essential that the formal conservation and plant breeding sector recognize farmers' capacity to maintain and develop local diversity. Linkages which recognize and build upon farmers' skills need to be built between the formal and local crop development sectors.

Local crop development

Landraces formed the basis for the development of plant breeding, long before Mendel's genetic discoveries influenced modern plant breeding. Farmers not only maintained a wide variety of landraces, but also continuously evaluated and improved their planting material and exchanged it with others, developing skills in experimentation. These processes lie at the very heart of agriculture and have supported its expansion into diverse environments.

The tradition of farmers developing and maintaining diversity can still be found in many areas of the Third World (Richards, 1985), where it has not been undermined by the expansion of modern varieties. Hardon and de Boef (this volume) define the concept of local crop development as the continuous and dynamic process of maintenance and adaptation of germ-plasm to the environment and to specific local household needs. Various authors have described and defined this process from a plant breeding and seed supply perspective (Berg et al., 1991; Groosman et al., 1991; Cromwell et al., 1993);

3

others have described local crop development from a conservation perspective (Keystone Center, 1991; Brush, 1991; 1992; UNEP, 1992; Friis-Hansen, 1993; Wood and Lenné, 1993). Hardon and de Boef argue that plant breeding, seed supply and conservation perspectives are combined at the local level in farmers' strategies of local crop development. While the specialized commodity-sector organization of modern plant breeding encourages the separation of conservation and plant breeding activities, at the farm level both these activities are an integral part of local crop development.

In the professional organization of the seed industry, the different elements of the seed supply system have been separated into disparate, specialized units such as genebanks, plant breeding institutes and companies, seed production farms, etc. (Cromwell *et al*, 1992). This system, which was developed in the United States and Europe early this century, has been copied in many developing countries (Groosman *et al.*, 1991; Cromwell *et al.*, 1992). Professionalization within the seed sector has forced an increasing uniformity in agriculture based on the use of 'high-yielding' or modern varieties. These varieties include single line varieties for self-pollinating crops, hybrids for cross-pollinating crops and single-clone varieties for vegetatively reproduced crops. They are selected for high-yield characteristics in a broad variety of environments and depend upon the use of external inputs and standardized cultural practices to achieve their potential.

The professional seed supply and plant breeding sectors are served by a system of legislative standards for seeds quality and variety protection. These standards have created for breeders criteria which implicitly reject diversity and the development of variety mixtures and synthetic varieties as new, more appropriate, options for plant breeding (Berg *et al.*, 1991). Commercial interests have also created a whole system of intellectual property rights and patenting rights which encourages standardization and uniformity. Rights in local plant genetic materials have been transferred from farmers to the professional seed supply systems. This has been achieved through a failure to recognize the role farmers have played in the development of landraces, the basic material of the professional seed sector (Fowler and Mooney, 1990).

In the modernization of agriculture in developing countries, seeds and varieties play an essential role. However, the seed sector has been isolated from the crop development initiatives of small farmers, and has primarily been influenced by commercial considerations, rather than the needs of the majority of farmers in marginal areas. There are noticeably few institutions concerned with developing appropriate, integrated conservation and crop improvement strategies for marginal areas. A precondition for developing such a strategy is to build linkages between crop breeding research programmes and farm-level crop development. This will require new methodological approaches and institutional innovations which involve farmers in the processes of technology development, and which seek to strengthen the innovative and experimenting traditions of rural communities.

Farmer participation in agricultural technology development

While there has been a significant adoption of modern varieties in the higher potential environments of the South (such as Mexico and south-east Asia), in more marginal environments, in mountainous, arid and fragile environments, the process of technology adoption has been problematic. Seed and input packages have often been resisted and rejected by farmers. Technologies which were tested on experimental stations, in conditions which did not reflect the prevailing small-scale farmer environment, often performed poorly under farmer management.

In the 1950s and 1960s agriculturalists argued that the failure of modern varieties to spread into marginal areas was rooted in the conservatism and backwardness of traditional farmers. It was assumed in the dominant *transfer of technology approach* of the day, that research station technologies were superior to farmer practices since they performed better on-station with standardized inputs. The process of technology generation would be slow but onward marching. New technology would first be adopted by a handful of progressive farmers and then gradually filter down to the smaller more backward farmers as they realized the benefits which accrued to early adopters. However, in due course, it became apparent that this was not occurring, and that the new technology did not spread beyond the confines of richer farmers and better endowed areas.

From transfer of technology to farming systems research

By the 1960s the beginnings of a critique of modernization theory in agriculture became manifest. A number of researchers argued that the failure of farmers to adopt technology was not the result of their conservatism, but emanated from inappropriate design of the technology for the farmers' ecological and socio-economic conditions (Wilde, 1968; Mellor, 1966). Researchers began to focus on the constraints influencing farmers' perceptions and strategies, and sought to understand the rationales underlying farmer practices. By the 1970s a number of social scientists began to argue that small-scale farmers in fragile environments devised strategies in which maximization of yield was only one objective among many. In these unpredictable environments, farmers were often more concerned with developing risk minimizing strategies, which ensured that they would get some yield no matter the unpredictability of climatic and other factors. Farmers often preferred hardy to high-yielding but vulnerable crops and often engaged in intercropping to minimize risk (Norman, 1974; Collinson, 1972).

These developments paved the way for the emergence of farming systems research (FSR) in the 1970s, an institutionalized form of multidisciplinary research in which social scientists were charged with developing research methodologies and approaches which would tailor international agricultural research to the needs of small-scale farmers (Hilderbrand, 1981; Byerlee *et al.*, 1982; Rhoades *et al.* 1987). FSR received support from the Consultative Group on International Agricultural Research (CGIAR), and multidisciplinary team research was incorporated into the international agricultural research centres. In FSR the social scientist emerged as a broker between farmers and agronomists, elaborating models of the small-scale farmer environment which could be incorporated into the development of appropriate technology designs, and devising mechanisms to ensure that feedback from farmers could be incorporated into technology testing programmes.

At the International Potato Center (CIP), social scientists were integrated into interdisciplinary teams. They complemented the work of the natural scientists, by unravelling farmers' perceptions of particular problems, and by showing the

5

effects of socio-economic and cultural factors on the perceptions of farmers. Scientists working on-station in Peru had identified storage losses to be a major problem. However, working within villages the social scientists found that small shrivelled potatoes were not considered a problem by farmers but had their uses. Careful research revealed that the main storage problems involved excessive sprouting of new potato varieties in traditional dark room storage. Eventually storage in diffuse light was identified as a solution to the problem, and on-farm testing with the participation of farmers led to the design of a technology which fitted into farmers' living conditions and was compatible with their needs. The technology was widely adopted by farmers. This experience led to the codification of a new research model, the *farmer-back-to-farmer* model, in which research is seen as a continuous process which is both interactive and iterative, and in which ongoing evaluation of technology by farmers plays an important role in influencing the process of technology development (Rhoades and Booth, 1982).

Co-opting farmers into technology testing?
This interactive model of agricultural technology development was further elaborated by a number of researchers. Chambers and Ghildyal (1985) developed the *farmer-first-and-last* model, in which small farmers could participate in defining the research agenda, in planning and prioritizing technological needs.

On-going research also revealed that farmers carried out their own experiments and innovations, and disseminated relevant innovations through networks (Box, 1984; Richards, 1985; McCorkle et al., 1988; Haverkort et al., 1991; Rhoades and Bebbington, 1991). This has led to attempts to incorporate farmers into programmes of technology development, involving them in on-farm testing and also in designing trials. During the late 1980s this recognition underlay an emerging conception of farmer participatory research (FPR) or participatory technology development (PTD). In place of the emphasis in farming systems research on modelling the small-farm environment and introducing replicative trials for technology testing, it was argued that more cost effective feedback could be established by allowing farmers to participate in technology testing (Matlon et al., 1984; Farrington and Martin, 1988; Amanor, 1990; Chambers et al., 1990). Some of these trials are researcher managed, but others have encouraged farmers to manage their own trials, and have recognized the fact that farmers often approach and conduct trials with aims and objectives that often differ from those of researchers. Farmer-managed trials have revealed criteria which farmers use in selecting and rejecting new varietal matter, such as early maturity, flavour, ease of weeding etc., which have often been overlooked by researchers (Ashby, 1987).

Richards (1987) has raised the problem that involving farmers in researcher trials may co-opt their own independent experimental traditions, and marginalize lines of experimentation which do not fit into the dominant precepts of researchers. Van der Ploeg (1990) has also argued that farmers' independent research traditions are built on disparate constructs that often differ from those of modern commercial agriculture. They constitute an 'art de la localité', a system of continually adapting and matching seeds and technologies to changing environmental conditions. This system is threatened by modern agricultural science which seeks to transform farmers into consumers of its technology, and to replace the dynamic processes of adaptive response to micro-environments, with the consumption of standardized inputs which transform and mask interaction with the environment. Thus, even more sensitive targeting of small-scale producers by agricultural science and participation in researcher generated programmes can run the risk of co-opting farmers into using researcher-generated technologies and modes of experimentation based on paradigms which undermine the basis of farmers' own adaptive research traditions.

Building farmers' innovatory capacity

A third approach is less concerned with co-opting farmers into pre-existing programmes of technology testing, and more with acting as a catalyst, facilitating farmers to diagnose their problems, to identify and prioritize possible solutions, and to experiment in pursuit of those solutions. Such programmes are often highly critical of existing technology options, and encourage local self reliance and diversification away from modern varieties (Tan, 1986).

While many participatory research programmes involve farmers in the testing of formal sector technologies, a few have been concerned with those of farmers. At the Khon Kaen University Farming Systems Research Project in Thailand, it was recognized that farmers' own innovations were relevant and could form a base for generating new technologies. Farmer innovations were screened for their adaptive potential in other areas and the innovators were used as trainers to extend their technologies. Farmer technologies included cropping systems such as groundnuts after rice (Jintrawet *et al.*, 1985), sesame before rice (Wilairat *et al.*, 1985), and small dairy management practices (Simaraks *et al.*, 1986).

Farmer knowledge and the environment

Other research approaches have been concerned with the impact of technology development and commercialization on local cropping systems and cultivars, the negative effects of pesticides, developing the potential of local cropping systems and cultivars, and low external input agriculture. Such approaches are apparent, for instance, in the Latin American Consortium on Agroecology and Development (CLADES), and the Information Centre for Low External Input and Sustainable Agriculture (ILEIA).

Recent concerns with the environment and natural resource management, and the recognition that technologies need to be adapted to specific environments and farming systems rather than uniform environments, have resulted in a growing emphasis on developing research approaches which build upon the capacities of farmers to engage in experimentation and adapt technologies to local needs. These approaches place a strong emphasis on the local knowledge of farmers, both as an entry point for building new research capacities, and as a relevant resource which often complements shortcomings in formal scientific approaches to interaction with the environment.

Limitations of farmer participatory research

Biggs (1989) has developed a typology of farmer participation in agricultural research. In *contractual* research the role of farmers is minimal, providing farm resources for the researcher to experiment with. *Consultative* research involves researchers soliciting farmers' problems and then developing solutions for the farmer. In *collaborative* research farmers participate in various stages of research. *Collegiate* research entails researchers working to strengthen the experimenting traditions of farmers. However, effective farmer participation is difficult to achieve since relations between farmers and researchers take place in the context of wider political relations in which farmers are marginalized and have poor access to resources and science.

The international division of agricultural research, in which the important applied research is carried out in international centres while developing country national agricultural research services (NARS) concern themselves with adapting the products of international research to national conditions, creates structures which are frequently indifferent and insensitive to the needs of small farmers (Biggs and Farrington, 1991). Since funding for NARS programmes in developing countries is often based on their linkages with the international agricultural research centres, they develop objectives which reflect the policies of international research centres rather than the perspectives of farmers. This

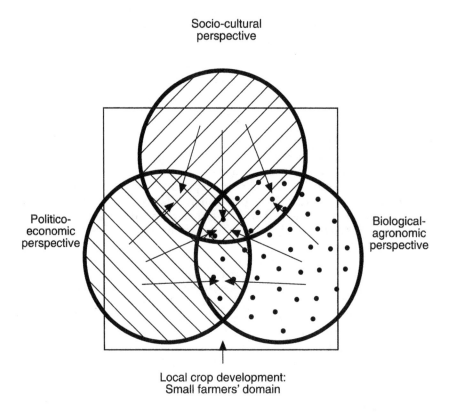

Socio-cultural
perspective

Politico-
economic
perspective

Biological-
agronomic
perspective

Local crop development:
Small farmers' domain

Figure A: *The relationship between small farmers and researchers in local crop development*

tends to uphold a central model of technology development and diffusion in which the generation of technology is viewed by agricultural scientists as the monopoly of international agricultural science.

Policies based on a central model of technology diffusion (Biggs, 1989; Farrington and Biggs, 1990) are at variance with a meaningful conceptualization of farmer participation in agricultural technology development, and a recognition of the validity of farmer experimentation. They marginalize locally-based research projects, involving groupings of farmers, local organizations, NGOs and researchers, operating outside the framework of research centres and extension services and outside the sphere of disseminating internationally-generated technology packages. A redefinition of the research system is needed in which innovation is recognized as emanating from multiple sources (Farrington and Biggs, 1990), and in which farmers are recognized as a resource community and not as consumers of agricultural technology packages (Röling, 1990). These conceptions need to be integrated into a policy framework which is concerned with the role of farmers in experimentation and technology generation, and with the implications of the present structure of agricultural research and research funding for the linkages between farmers, the environment, on-going locality-based research (which may involve NGOs), and agricultural science.

In local crop development, it is the concept of collegiate research which needs to be upheld to promote a relevant tradition of conservation and development of local genetic resources. For this to be achieved, the institutional context and policy framework of research has to be analysed in order to understand (and then reverse) the processes that lead to the marginalization of small farmers. Methodologies, approaches and new institutional innovations need to be developed which build linkages between researchers and farmers, and strengthen the capacities of farmers to articulate their requirements and develop their own traditions of experimentation and innovation.

Building bridges between farmers and researchers

In recent years, researchers, development workers and NGOs have been developing methodologies and organizational forms to support processes of local crop development and experimentation, to strengthen the role of farmers in setting and prioritizing their own research agenda, and to build farmers' capacities to represent and articulate their interests.

Supporting processes of local crop development

Various strategies and methodologies are being experimented with as a means of supporting existing processes of local crop development. One approach is to provide farmers with a 'basket of choices', expanding the options from which they can engage in experimentation and develop initiatives. The Ethiopian Plant Genetic Resource Centre, for instance, provides farmers with elite germ-plasm to further their experimentation in crop development. Farmers' landraces are preserved in the genebank, and are accessible to them at any time that they wish to return to their seeds (Mekhbib and Worede, Part two).

Other approaches focus on building research capacities in local organizations. In the Andes, seed fairs are important avenues through which farmers exchange local seeds and knowledge about crops and varieties. One group of researchers

9

is attempting to consolidate these conservation and crop development traditions by organizing seed competitions within the fairs. These competitions serve to revalidate both local knowledge and the contribution that farmers have made to the development of seeds. This has generated much interest among peasant farmers and acted as a forum for encouraging the maintenance of seed diversity and local crop development (Tapia and Rosas, Part three).

Some national and international research programmes are also involving local communities in co-ordinating and managing seed banks. In these programmes, farmers act as curators, collecting, identifying and preserving species. The research centres provide material incentives and equipment to enable the farmers to develop their potentials (Prain, Part three).

NGOs are also supporting local crop development activities. Strengthening farmer and community management and co-ordination of seed systems is an important component of their work. Projects have included: community seed banks (such as those established by ACORD in Mali, and by Oxfam in Sudan); inter-community seed exchanges; and local seed multiplication and distribution (for instance, in Bangladesh, Nepal, Ecuador and various countries in Africa). Cromwell et al. (1993) note, however, that agencies have been quicker to set up parallel systems (which they identify as local seed multiplication and distribution) than to work with existing systems (which they call farmer seed systems).

Other interventions by NGOs aim to change the approaches and methods of formal sector researchers and institutions, developing collaborative ventures which promote more sympathetic interactions with farmers. Offering training to formal sector development workers in participatory research methods is one way of achieving this. Other NGOs have provided courses on themes such as indigenous seed collection for government researchers and extensionists (Arum, 1993). Joint farmer-researcher trials can also provide the opportunity for practical collaboration. In Northern Ghana, for example, NGOs have organized farmers into groups which interact with formal sector researchers, participating in field trials and developing feedback to researchers (Kolbilla, 1993). They also facilitate communication by acting as an interpreter of language, methodological approaches and values between the two knowledge systems. The Association of Church Development Projects, operating in the same area, is collaborating with government researchers and extensionists to translate findings from formal sector research and local knowledge into practical extension messages (Alebikiya, 1993).

Apart from increasing mutual understanding, projects which link formal sector and local crop development systems can lead to the sharing of material, and the production of better varieties. The Institut des Sciences Agronomiques du Rwanda (ISAR) and the International Center for Tropical Agriculture (CIAT) have progressed beyond merely inviting farmers to evaluate their field trials to a proactive relationship in which Rwanda farmer-seed-experts, mainly women, are now evaluating varietal trials at an earlier stage in the breeding process, sharing expertise directly with station breeders, pathologists and agronomists, and selecting varieties to test directly in their own fields (Sperling, 1988). Station researchers have been sceptical about farmers' ability to judge seeds which they had only seen on-station, but farmer-selected varieties have outperformed the control variety.

Involving farmers in the evaluation of new technologies and understanding their criteria for assessing varieties is an important part of fostering respect for, and development of local knowledge. Techniques for eliciting assessments from non-literate farmers have been developed based on traditional African board games, with farmers using stones and seeds to indicate their preferences and to rank characteristics, technologies, expected returns etc. (Barker in Brokensha *et al*, 1980). Such techniques have also been used successfully in agroforestry programmes, where evaluation is complicated by the combination of several different components and multiple objectives (Scherr and Muller in ILEIA, 1991).

Strengthening local capacity to plan and manage development

Some development organizations are working to promote institutional developments which can strengthen the position of farmers and thereby increase the responsiveness and accountability of professional plant breeding. Participatory methods underpin many of the efforts to improve the capacity of people within rural communities, particularly powerless groups, to plan, evaluate, co-ordinate, manage and negotiate for control of development activities.

Increasing people's capacity to plan their own development underlies the activities of many NGOs and grassroots' organizations. The Myrada project in South India described by Mascarenhas (Part three) uses participatory rural appraisal (PRA) methods, such as drawing simple maps with villagers. These methods help them and outsiders understand the dynamics and envisaged development needs of their area and to formulate development priorities and strategies. The methodology has also been introduced into government planning organizations. A major achievement is that watershed maps are now being prepared not in offices but out in the villages with the people.

Farmer representation and empowerment

Organizing local people into effective groups is seen as the key to improving the position of the rural poor by strengthening the capacity of local organizations to negotiate with outside agencies and make development workers aware of their needs (Bebbington, 1991). Farmer groups and networks have been established in a number of countries in the South, some having representation in national research systems and government policy bodies, or internationally, through the International Federation of Agricultural Producers (IFAP).

NGOs are working to improve farmers' access to seeds at national and regional levels. In the Philippines, a seed exchange network has been established which accesses rice varieties from over 100 NGOs and farmers' organizations as well as research institutions. The Rural Advancement Foundation International (RAFI) and African Seeds of Survival provide practical support for community level conservation and utilization. This is combined with awareness-raising programmes about access to plant genetic resources and other issues across Africa and in the North (Cromwell *et al.* 1993).

Policy and the wider political and economic environment

The impact of methods and institutional linkages on local crop development is, like all development activities, dependent on national and international policy and the prevailing political and economic environment. Some NGOs have incorporated the consciousness-raising of rural people, and the agents with whom they come into contact, into their programmes. However, local initiatives are continually challenged by big business seeking to expand its sales of high-yielding varieties, fertilizers, chemical and other inputs developed under the protection of plant breeders' rights and patents. A number of NGOs and other organizations believe that the maintenance of sustainable agricultural systems on small farms depends on raising global awareness of the dangers of declining genetic diversity and the inequity of breeders' rights which ignore farmers' rights in local genetic resources (Cromwell *et al.* 1993). Some are now highly active in direct lobbying for reform in international and national seed and patenting policy.

International and national seed legislation and policy have impacts on local crop development at various levels. Regulations on organizations permitted to undertake breeding work and procedures for releasing new varieties frequently appear to be based on the need to preserve the monopoly of seed breeders, while the long time periods stipulated for testing can severely restrict the availability of introduced material. The Zimbabwe Seeds Action Network is a group of NGOs working in the collection and development of local varieties of millet and sorghum, which have hitherto been largely ignored by the formal sector and are not subject to the same controls as maize. Members of the Network are also working on open-pollinated varieties of maize and have opened a dialogue with the Ministry of Agriculture on testing.

In many countries, however, constructive engagement is conditioned by what many NGOs and plant breeders feel to be the biggest threat to local crop development: patenting laws for life forms. The Alternative Technologies Project (PTA) in Brazil, a network of NGOs collaborating with farmers and researchers to maintain and reintroduce local varieties of maize which are being displaced by modern varieties, is one such project which will be unable to continue its present system of free access to germ-plasm maintained by farmers if new patenting laws come into force (Cordeiro, Part four).

At the international level, campaigns for a favourable political environment in which farmers' rights in seeds will be respected are directed at policy makers and the wider public. The International Coalition for Development Action Seeds Campaign and Genetic Resources Action International (GRAIN) have sought to stop national legislation granting monopoly control over plant genetic resources to seed companies, and are promoting the adoption of international agreements on exchange and conservation of genetic resources. GRAIN's work involves raising awareness among policy makers and the public (mainly in the North) about the causes and implications of genetic erosion. RAFI is working with regional NGO networks to make the issues surrounding international plant genetic resources widely understood and to act as an early warning system for developments that might adversely affect farmers' access to them. These NGOs have been active in promoting the FAO International Undertaking on Plant Genetic Resources and the concept of Farmers' Rights (Cromwell *et al.*, 1993).

Apart from what they see as the political influence wielded by agribusiness over patenting policy and research, NGO lobbyists identify economic power as the biggest threat facing producers of local crops. By a sometimes subtle but perpetual process of promoting western-style values and products, systems of local knowledge, crops and values are being eroded. Southern NGOs are acting to counter these powers by championing indigenous values and systems (Opole, Part four). Others are supporting the preservation of local plant and genetic resources through encouraging their productive and sustainable utilization (Mbewe, Part three).

Many supporters of local crop development have found that strengthening local capacity to lobby for policy changes is the soundest way of creating a self-sustaining process driven by local people. Alongside research, education and dissemination activities, NGOs are working to empower local organizations and to create national social movements to press for change (Korten, 1990; Clark,1991; Farrington and Bebbington, 1993).

New challenges for crop research

The value and relevance of farmer knowledge of crop genetic resources is being discovered in a range of disciplines and institutions. But recent interest in local knowledge of genetic resources needs to be treated with caution since, in this age of biotechnology, control over diversity in plant genetic resources offers the promise of considerable wealth and economic power to the agribusiness world. This is reflected in the recent rush to establish patents over life forms. In this struggle the odds are very much stacked against small farmers. Through control of world legal, market and political knowledge, agribusiness interests are likely to be able to use research into local knowledge for narrow commercial interests, rather than for the long-term development and environmental needs of marginalized farmers. There is the risk that discovery, while potentially progressive, can precede exploitation.

In the wider context, the great challenge for local crop development is to find means through which local knowledge of crop genetic resources can be validated and linked to world science, while at the same time ensuring that control of that knowledge remains within the community.

An adequate response to this challenge requires that, beyond developing methodologies and institutional approaches for building linkages between farmers and researchers, research addresses the policy dimension of crop development. This requires the elaboration of a critique of the institutional and commercial settings of research, a critique that has the potential to influence policy frameworks and promote change and reform in such areas as legislation, intellectual property rights, and agricultural research management. This critique must also create pressures for technology institutions to become more transparent and responsive to the needs of the majority of farmers. At the same time, local organization needs to be strengthened in order to increase the capacity of farmers to manage local development, to analyse the wider environment in which their crop development activities unfold, and to exercise pressure over institutions and policy makers (Farrington and Bebbington, 1993; Carroll, 1992). These are challenges which can only be met when socio-cultural, biological and political economic perspectives are combined consistently and logically.

PART ONE

UNDERSTANDING FARMERS' KNOWLEDGE

Introduction

Kojo Amanor

Early studies on farmers' knowledge systems tended to stress their utility for agricultural development, to focus on the rationality of indigenous classification systems, and to illustrate the sound adaptive practices of farmers. In recent years research has increasingly become concerned with the relationship between local knowledge and science and its implications for the independent experimental traditions of farmers and for agricultural development. Researchers are increasingly examining the context in which knowledge is generated, the relationship between technical knowledge, social institutions, socio-economic and political relations and the marginalization of small-scale farmers.

These concerns are reflected in the papers in this section, all of which examine the role of local knowledge in the development of technology and its implications for agricultural research and development approaches. The papers also reflect different ways of looking at local knowledge. Some of them are concerned with the interaction between farmers and researchers in conventional research and development settings (Botchway, Dusseldorp and Box, Millar), and others focus on description and analysis of farming systems and the implications of local practices for wider agricultural research policy frameworks (Amanor, Longley and Richards). However they all point to the complexity of local knowledge systems and reflect an increasing focus on local knowledge as process. They are concerned with the ways agricultural knowledge is embedded in social, cultural and ecological systems. This includes the context in which agricultural knowledge develops, and the implications of the marginalization of rural communities and particular groups within communities for the development of knowledge at the local level and within formal research systems.

Dusseldorp and Box examine the interaction between farmer and scientist knowledge systems in adaptive research focusing on communication. They address the implications of differences in perceptions, worldviews, and approaches to knowledge for interactions between farmers and scientists. They stress that all knowledge is relative and must be understood within its social and cultural context. Scientific knowledge is constrained by its specificity, its commodity and discipline focus and its standardized research procedures. It is argued that to be more responsive to the needs of farmers, agricultural research has to create a more interdisciplinary base in which social scientists play important roles in eliciting the perceptions of farmers and understanding the significance of local knowledge within its social and cultural background. Social scientists also need to develop analysis of the constraints within formal research programmes arising from their own social and cultural contexts.

Botchway vividly describes how the failure of the Weija Irrigation Project to take farmers perceptions and strategies into consideration adversely affected project activities. He shows how the different socio-economic and cultural contexts in which farmers and development projects operate can lead to clashing objectives and strategies. The project operated within a narrow cultural focus of optimizing yields through use of high inputs and control over the marketing of farmers' outputs, without taking farmers' strategies and perceptions into consideration. Failure to understand the social system, the household economy and gender division of labour resulted in women being alienated from the project and unrealistic labour demands being made on farmers. The project is now attempting to rectify these problems by incorporating an awareness of local perceptions, strategies, and social and cultural organizations into project design. In his analysis of local knowledge, Botchway examines the relationship between local perceptions of farming activities and development projects, economic strategies rooted in the household, other rural institutions, and the village economy, and the problems of allocating and managing labour resources.

Amanor examines the relationship between farming strategies, labour constraints, and environmental degradation and the implications for the wider research system. He focuses on agroforestry systems and farmers' responses to degradation and problems in their bush-fallowing system in the light of historical experience. Farmers' knowledge systems are seen as dynamic: continuously interacting with the environment and changing as new problems are encountered. The study focuses on farmers' adaptation and responses to changing environments by contrasting their experiences in localities characterized by differing rates of degradation through time. Knowledge is seen as arising from attempts to wrestle with problems rather than from familiarity with stable systems of traditional land management. The range of strategies, innovations and experiences vary considerably in different microenvironments. Interaction with the environment has given farmers insights into natural resource management which often go beyond the limits of formal scientific understanding and methods. Given the experimental abilities of farmers and the constraints of the formal agricultural system in adapting technologies to diverse environments, it is suggested that agricultural policy frameworks should attempt to strengthen the adaptive and experimental strategies of farmers as a basis for the development of local-level research and technology options.

Millar is also concerned with the interface between formal research systems and farmers' experimentation. However, he focuses on the relationship between farmers' technical knowledge and their worldviews or 'cosmovisions'. He argues that these worldviews and the institutions and social relations connected with them play an important role in defining the adoption or rejection of technologies, and the modes through which farmers carry out their own experimentation. While these world views may appear to be irrational to scientists, the actual processes of experimentation that farmers engage in follow logical procedures similar to formal science. He argues for the development of a more holistic approach to research, which allows farmers to process technology in accordance with their worldviews and enables them a greater role in the processes of research with the aims of creating more flexible technological options. This requires that science becomes more aware of the cultural context in which agricultural activities occur, and attempts to work within the precepts of the farmers. However his analysis does not extend to the socio-economic and political basis in which worldviews

18

are rooted, the differing ideologies of different social groups, and factors resulting in changing worldviews.

Longley and Richards examine the cultural and social context in which innovation and experimentation take place. They contrast two experimental traditions in Sierra Leone, and point to how factors concerned with different social structures, gender relations, and cultural perceptions of stratification and ecology affect experimentation with rice varieties. They argue that social and cultural factors are as important as technical considerations in influencing the course of agricultural experimentation, within and between different communities. While only the paper of Longley and Richards deals specifically with crop genetic resources, many of the concerns with the processes of experimentation in local crop development are reflected in section two.

The failure of formal research institutions to incorporate an understanding of local knowledge and its context into policy frameworks, planning and the implementation of projects often has negative impacts on the development process. Major issues arising out of these studies on local knowledge include the search for new institutional and organizational forms, and research approaches and methodologies which will strengthen farmers' ability to place their needs on the development agenda while strengthening their own autonomous traditions of innovation and production.

Local and scientific knowledge: developing a dialogue

Dirk van Dusseldorp and Louk Box

This paper addresses problems in the linkages and articulation between local and scientific knowledge systems and networks. It is argued that dialogue is essential to the interaction between the two systems and problems which impede dialogue are identified. These relate to differences in worldviews, vocabularies, values, methods of accessing and approaching knowledge and differences in risk assessment. Social scientists are seen as playing an important role as brokers between the two systems, facilitating dialogue. They can examine how all technical knowledge is embedded within social and cultural systems, enabling the limitations and approaches of both knowledge systems to be perceived and new forms of interdisciplinary adaptive research to emerge.

The role of local knowledge in agricultural knowledge has been overlooked. Although social scientists have pointed this out, agri-scientists have only recently shown interest in local knowledge. Sometimes social scientists have gone to the other extreme and emphasized local knowledge to the extent that the role of the knowledge of research institutes in technology development is neglected. We argue that the two systems of knowledge are equal and complementary, but with quite different attributes and modes of social organization.

The meeting of these two knowledge networks and the broadening of the disciplinary base of research institutes could result in more relevant knowledge and approaches to sustainable agricultural knowledge. The fruits of this meeting depends on the quality of dialogue between the actors. This is an essential prerequisite for a successful combination of elements from different knowledge systems. This paper addresses problems concerned with the linkages of the two systems and examines some of the obstacles which impede communication and modes of promoting dialogue.

Communication, dialogue and rationality

Communication is the exchange of symbolic information. This information is partly, consciously or unconsciously, given, received and interpreted (Oomkes, 1986) between people who are in one way or another aware of each other. The way communication takes place is culturally determined. In different cultures languages and symbols with disparate meanings and associations are used. The way communication takes place also differs between ethnic communities (Eisstadt, 1955; Hyman, 1966).

Dialogue is important in interpersonal communications between actors of equal standing, where there is the possibility of exchanging arguments and contra-arguments. It involves constant feedback in order to find out if the actors have understood the information they have exchanged. This makes it possible to arrive at a complete mutual understanding and possible consensus (Reimann, 1974). This contrasts with 'Socratic dialogue' (Popper, 1968), where the main speaker has the superior intellect and the other partners are only asking questions for clarification. Yet it is 'Socratic dialogue' that often characterizes the communications between extensionists and farmers.

Communication requires empathy. This is the ability to project oneself imaginatively in another position (Theodorson and Theodorson, 1969:130). Empathy is important because in the knowledge systems of scientists and local people instrumental and traditional rationality play a role (Weber, 1980). In instrumental rationality, where objects and persons are used as relatively efficient instruments or means for attaining one's own ends, it can be expected that both parties will understand each other. However, in the case of traditional rationality, that is largely determined by the ingrained habits of the participants, empathy becomes important, because this can lead to a different understanding of causalities underlying the changes in their environment (van Dusseldorp, 1992).

Dialogue in agricultural knowledge systems: different perspectives

In the exchange of agricultural knowledge, a crucial issue is the mode of communication between cultivators and scientists. They have fundamentally different cultural backgrounds and symbolic systems, different socio-economic positions, and a different appreciation of risk. They have to exchange information which is difficult to fit in their respective worldviews and knowledge systems. A real dialogue is only possible when in the scientific knowledge system an intimate relationship has been developed between contributing disciplines. Similarly, a sensible dialogue can only take place when effective knowledge networks link individual cultivators, allowing them to exchange experiences and information.

In the development of agricultural science new ideas are frequently discarded because they do not correspond to prevailing conditions. Without the establishment of a sound foundation and conducive environment for dialogue, the prerequisite conditions which are needed for innovations to be sustained are often overlooked, leading to the failure of technology adoption. It is important to understand the problems in bringing together and merging the information available in local and scientific knowledge systems, since this enables the possible pitfalls to be identified and new approaches to be developed. These problems are related to differences in worldviews, vocabularies, values, approaches to knowledge, risk assessment strategies, and methodologies.

Worldviews: voluntaristic versus adaptive
Cultivators and scientists belong to different knowledge networks, and they generally have different cultural backgrounds. Culture, in the classical definition of Tylor (1871), is 'the complex whole which includes knowledge, belief, art, morals, law, custom, and any other capabilities and habits acquired by man as a member of society' (Theodorson and Theodorson, 1969:95). Thus, there are likely to be differences in the world views of scientists and cultivators. Those

whose action is mainly embedded in a scientific knowledge system (researchers, technologists, teachers and extensionists) assume that people have the potential to understand the processes of nature, to express underlying causalities in theory, and to harness theory to manipulate the environment. In this *voluntaristic* worldview, humans and not supernatural beings make the world. Those who belong to local knowledge networks (cultivators, small traders) generally share the world view that a large part of their environment is controlled by powers beyond their reach and use rituals and offerings to please these powers. They accommodate changes in their environment and therefore have an *adaptive* worldview.

These contrasting world views lead to different causal explanations of what happens in the environment. For example, in Sarawak, a Dajak farmer saw that his rice was not growing well. He assumed that spirits were sucking away its growing powers and slaughtered a chicken to appease them. An extension officer came to the field and explained to the farmers that it was not spirits but insects (*jassiden*) which were attacking his rice and that these could be killed with DDT. This incident took place over ten years ago, when DDT was widely used and its toxic properties not well known. Although both responses may be thought to be inappropriate now, within the context of their time and environment both actors were operating rationally. Their behaviour was consistent within their system of logical thought.

For linkages to develop between the two knowledge systems, dialogue is needed in which both farmers and development workers are willing to listen to each other as equal partners in agricultural change. For dialogue to develop it is necessary that the actors are aware of the differences in their world view, and develop an empathy to understand the reasoning of the other party and their view of causality. Research carried out in the Dominican Republic revealed widely differing views on problems in cassava and rice production between cultivators, extensionists, researchers and scientists, with greatest differences occurring between cultivators and scientists. It was only when the authors revealed these differences to the groups involved that they became fully aware of them. Social scientists can stimulate dialogue by pointing to existing mergers between previously different world views (Box and van Dusseldorp, 1992).

Vocabularies: local versus academic

Before dialogue can take place, development workers need to have an intimate knowledge of the local dialect and its vocabulary. In cassava research in the Dominican Republic we found that many scientists did not know the local names of varieties, or local names of particular factors leading to harvest failure. This was remedied by extensive interviewing of key informants and the collection of vocabularies concerned with varieties and diseases. An understanding of the causalities, as seen in the local knowledge system, requires comprehension of the language in which these terms are embedded. Since agro-biologists lack the training and inclination to carry out this kind of work, it is best undertaken by social scientists, as demonstrated by the work of Rhoades at CIP (Rhoades, 1982; Rhoades *et al.*, 1987;). Through case study approaches and key informant interviewing, data can be analysed by social scientists to enable agro-biological scientists to participate in adaptive research.

Values: specificity versus universality

Local knowledge is embedded in a specific geographical and cultural context. Scientific knowledge is embedded in a value system aiming at universality. These differences are evident but should not be exaggerated. Elements of local knowledge and the rules on which technologies are constructed may be embedded in far larger systems, transcending geographical and cultural specificity. Thus farmers in both the Dominican Republic and Asia use lunar positions for developing rules for cultivation. The universality of science may also have its limits as shown by Kuhn's classical work on paradigms (Kuhn, 1970). As a means of facilitating dialogue between the two systems, social scientists can draw general elements from local knowledge and point out the specificity of scientific knowledge.

Knowledge: holistic versus segmented

Local knowledge is holistic. Cultivators and their families seldom isolate one aspect of crop performance, such as higher production potential. Other important factors include the duration needed for crop processing, taste, and the fit of specific crop labour requirements into the social and ritual calendar. In contrast, scientific knowledge is characterized by specialization and segmentation. This limits the opportunities for productive dialogue between scientists from different disciplines and between agro-biological scientists and cultivators. Dialogue is thus dependent on the creation of multidisciplinary research and co-operation of different disciplines holding disparate paradigms and methodologies. In interdisciplinary research, however, there is often competition between disciplines. In several international research centres, sociologists and cultural anthropologists have been able to cooperate with agro-biological scientists, but have had considerable problems with the economists, who saw them as intruders in a field in which until recently they had a monopoly (Box and van Dusseldorp, 1992). These problems can only be solved when there are changes in the education system which produces scientists. This does not mean universities should focus on producing generalists instead of specialists, but that they ought to encourage the development of methodologically versatile specialists who are able and willing to co-operate with other disciplines. In experiences in interdisciplinary research in both the Philippines and the Dominican Republic, we found it more difficult to bridge the gap between individual disciplines than between researchers and cultivators. It requires great methodological virtuosity to undertake transdisciplinary research. We suggest that training programmes in adaptive agricultural research be institutionalized in universities.

Differences in risk assessment

It is important for scientists to understand that farmers assess risks in a different way from them. These differences are rooted in contrasting world views, interpretations of causalities and experiences of economic security. Farmers' assessments of risk may include supernatural sanctions, a holistic view of the environment, an awareness of the consequences of particular interventions for the household and community, and in the context of economic insecurity an aversion to taking risks which can endanger the existence of the household. The assessment of risk is an important factor which will influence farmers' decisions on the viability of implementing new elements from the scientific knowledge system.

Methodologies: adaptive versus directive

In mainstream agricultural research problem-identification, research design and data analysis are done on-station by senior researchers. Problem identification may be supported by field visits, but this is not systematically integrated into research. Research design conforms to standard rules and dogmas of valid and reliable techniques. Data analysis is performed following set rules which facilitate comparisons from which conclusions can be drawn. Each of these procedures systematically excludes cultivators from making an input in research. New research methodologies need to be found to foster dialogue in each of the research phases whenever it can be productive. For example, Box (1989) has argued that adaptive research methodologies can be effective and efficient. He describes a methodology consisting of the following stages:

o   Identification of the main actors and interests, including formal organizations (farmers' associations, land reform co-operatives, credit and extension institutions).

o   Intensive case studies among informants who are actively involved in agricultural experimentation and are capable and willing to share their knowledge.

o   Adaptive trials both on-station and on-farm.

o   Representative sample survey among cultivator populations, verifying quantitative relations among variables identified in the case studies or the adaptive trials.

o   Systematic reporting in public meetings to cultivator groupings and other formal organizations mentioned in stage 1 to check on results in stages 2,3 and 4.

Bridging social strata for dialogue

Cultivators, extensionists and scientists generally come from different social classes. The relationship between government officials, including extensionists and researchers, and farmers is problematic. Stereotypes on both sides prevent open dialogue.

Farmers often think government officials make their lives more difficult, by imposing restrictions on them, such as sanctions against cutting down trees. They are seen as representatives of oppressive and often corrupt institutions. They are unwilling to visit cultivator's fields and when they do they provide services that are not free as the cultivators had often presupposed. Cultivators have also had bad experiences with the information imparted to them, because the messages communicated were either too general and not adapted to specific local conditions, did not fit into the activities of the household, or were completely inappropriate. In the Dominican Republic, we found that cultivators did not turn to extensionists when they experienced problems in their cassava fields. They would first take their problems to neighbours, friends or family, followed by traders. Only a small percentage indicated they would take their problems to extensionists.

Government officials generally see cultivators as ignorant and in need of instruction on proper farm management. When they do not follow instructions, they are considered stupid, traditional and lazy. These officials are convinced of the supremacy of scientific knowledge, consider that they automatically know more than the cultivator, and look down on local knowledge. Field officers are in the lowest echelons of their organizations. An important part of their status is derived from the supposition that they know more than cultivators and thus have a superior position. Most of them are so poorly paid that they take additional jobs or use their position to earn something extra. They often fail to build relations of trust with cultivators because they are regularly transferred. In some countries rules exist to prevent officers being posted to the areas from which they originate.

Scientists and extensionists will not become open-minded towards cultivators when they have to operate in hierarchical, administrative structures. Changes in organizational structure, procedures and personnel management, and the provision of transport to make regular field trips possible are required. Structural adjustment programmes which, however, nearly always have adverse effects on the allocation of resources to the lower echelons of government and rural areas, make these reforms difficult for many countries to achieve.

Lately non-profit Non-Government Organizations (NGOs) have been seen as excellent vehicles that can function as intermediaries between government organizations and the local population. They are small and thus beautiful. They are flexible and can therefore adjust themselves easily to local circumstances. Their workers are idealistic, devoted, not corrupt and willing to stay in the field with the cultivators for long periods of time. Therefore NGOs are capable of doing the same job better and cheaper than governmental organizations. But the characteristics of NGOs, mentioned above, bring with them also some potential problems if they are going to perform a role in the dialogue between local and scientific knowledge. Idealism and devotion are essential but professionalism is required at the same time and the latter was sometimes missing. Recently NGOs are giving more attention to the professionalism of their staff but this increases the costs of their operations. The way NGOs are working is labour intensive, time consuming and area specific. The results of their work has in most cases only limited recommendation domains. With the expected rise of salaries it is important that special attention is given to the cost-effectiveness of their operations. A more serious problem that can arise is that idealism can lead to hobbyism and dogmatism. Sometimes NGOs are so convinced that their approach and message is the true one and they are not very open to criticism. This can make them poor partners and go-betweens in the dialogue between local and scientific knowledge. But when NGOs are aware of their potential biases they can fulfil an important role in the exchange and integration of knowledge coming from different social systems.

Dialogue and culture conflict

A dialogue between actors engaged in local and scientific knowledge networks has consequences which go beyond improvement in agricultural practices, and may affect the world view of cultivators. They may realize that the influence of the supernatural on their environment is limited. This in turn may undermine the world view embedding their norms and values and lead to anomie, a state of normlessness or of conflicting values in society (Durkheim in Mann, 1983:13). Stimulating the dialogue between local and scientific knowledge networks will speed up this process.

A comparable process may occur among scientists, where they may come to realize they are not the trusted wizards they thought they were. Within the agricultural sciences a new balance needs to be struck, on the basis of a realization that the specific ecological context in which technologies are applied is significant. Within the agricultural sciences the most successful innovations can frequently be traced back to a dialogue between scientists and agriculturalists. This process has been largely hidden and does not appear in academic publications or applications for patents. It is hidden behind the curtains of institutionalized exchange between farmers and scientists in countries with a track-record of agri-technological innovation, such as the US, France, and the Netherlands.

Academic subcultures are resilient and perhaps as traditional as those of cultivators in the tropics. Changing them through dialogue with cultivators implies a certain degree of culture-conflict and normlessness. Potential problems of anomie should not stop efforts to continue the dialogue between local and scientific technologists. However one should be aware that this dialogue needs to be placed in a larger cultural context.

Concluding remarks

The gap between local and scientific agricultural knowledge is real and has been noted in the literature ever since the agricultural sciences emerged. In industrial agriculture, ways have been found to bridge the gap. In many developing countries, particularly in Africa, the social distance between scientists and cultivators is much larger than in industrialized countries. The need for dialogue is great in those situations.

In this paper we have argued that such dialogue needs to be structured: it will not come about by itself. Research techniques which provoke dialogue could be employed. But this is only a small, and possibly insignificant part to a solution. Non-governmental organizations may in certain cases provide solutions. Neither of these, however, can be sufficient.

Cultivator interest groups, capable of expressing themselves and of imposing their priorities on researchers, are likely to bring more permanent solutions. Only if local knowledge is empowered through institutional solutions, is a real dialogue likely to take place. This means that plant breeders will have to be creative in devising solutions to recognize the value of local knowledge, be it through communal cultivators' rights, patenting systems or in any other way which empowers local cultivators in their dialogue with agricultural scientists and in their dealings with other relevant pressure groups.

Implications of farming and household strategies for the organization of a development project

Jack Botchway[*]

The early formulation of programmes in the Weija Irrigation Project depended on technical considerations which did not evaluate problems within the local farming system. This resulted in serious problems rooted in conflicting demands on labour, unrealistic assessments of the availability of household labour, and alienation of female labour from irrigated plots. The paper examines the negative impacts which arose from this failure to take farmers' perceptions and strategies into consideration and describes new approaches developing in the project, which attempt to root research designs and planning in farmers' experiences and in the fabric of rural life.

In the 1960s and 1970s development projects failed to incorporate important aspects of local perceptions into their working procedures. The foundations of productive initiatives which had been laid long ago in rural communities and in local knowledge systems were ignored and replaced by technical expertise based on foreign concepts. The rejection of the validity of local knowledge has resulted in a large number of blunders which have led to the collapse of many projects.

The experiences of the Weija Irrigation Project point to the fact that where there is no understanding of the cognitive maps of rural people, project success becomes elusive and difficult to achieve. This paper shows how the study and application of local knowledge can contribute to the management of development projects. It focuses on problems of labour organization, farming strategies, and their impact on irrigated vegetable farming. It closes with an exposition on the organization of credit and how to successfully recover loans from farmers using novel techniques based on local knowledge.

Rationale of the project

During the early 1980s an irrigation project was organized by the government of Ghana on the Weija Lake, 25km to the west of Accra, using overhead sprinkler pipes. It was expected that by supplying food consumers directly in Accra, prices would be kept down and consumption would increase. Two hundred and twenty-five hectares have been developed as the first of the six

---

[*] Jack Botchway dedicates his paper to Issah Shaibu Agbo, an uneducated farmer of Tuba village who was one of the victims kicked out of the project in 1987 and who in 1992 received a Ghana government award during the Regional Farmers' day as an Indigenous Researcher.

phases of development of the project. The Weija Irrigation Project Feasibility Report (1976) identified land utilization in the project area to be well below capacity. It claimed that, in any given year, less than 30 per cent of the area was under crops and cropping intensity on the cultivated land was low. The introduction of irrigation was viewed as a means of promoting agricultural production through expanding the overall area under cultivation, increasing cropping intensity and raising crop yield per unit area. The main focus of attention was on intensification of farm production through irrigation along with new cropping patterns, improved agricultural techniques and high-yielding varieties.

The activities were organized around the formation of farmer co-operatives. Farmers jointly rented the land as tenants from the irrigation project. With the exception of labour, all other inputs for production were supplied to farmers through the project Extension Unit. This input scheme was developed alongside a supervised credit programme with detailed farm management plans which indicated how and when the inputs were to be used. The Agronomy/Extension Unit devised the cropping and planting programmes which were followed each year and the work for farmers was divided into two main periods: January to June and July to December. Without consulting the farmers the project authorities specified thirteen different crops which farmers were to cultivate in each period. The crops were tomatoes, okra, sweet potato, watermelon, cabbages, carrots, cucumber, cowpeas, onions, groundnuts, local garden eggs, aubergines and carrots.

This top-down approach to project implementation yielded poor results. This was not because the farmers were grossly inefficient and uninterested in new farming techniques, but because the intervention strategies failed to analyse adequately the farming system operating in the area and to incorporate data concerned with it into the project design. In short, the strategies and perceptions of the local farmers were grossly disregarded.

Farming systems

The main economic activities in the five villages surrounding the project are crop farming, cattle rearing, hunting, fishing, fish smoking and small-scale trading. The traditional farming system is based on rainfed agriculture. The rainfall follows the bimodal pattern characteristic of southern Ghana where the major rainy season peaks in May/June and the minor rainy season in September/October. During the major rainy season, cassava, hot pepper and maize are the main crops planted. Tractors are commonly hired for ploughing and harrowing, while planting, weeding and harvesting are carried out manually (Weija Irrigation Project, 1976). Tomato, one of the main cash crops in the area, is intensively cropped during the minor rainy season, but high humidity and temperatures experienced in the month of November, generally predispose it to heavy fungal attack and also cause severe flower drop. Staggered planting is therefore practised to spread the risk of crop failure while enabling farmers to respond to the high price tomato attracts just before the month of December and during the following dry season.

Most farmers who work on the project lands also maintain their traditional bush plots. In 1987, for instance, 96 per cent of project farmers had bush plots.

Farmers on the average had three bush plots in one farming season (Botchway, 1988) and they tenaciously hold on to their bush plots because they are much cheaper to maintain, the land is easily acquired and because there are no water costs for rainfed plots.

The majority of extensionists on the project in 1987 wanted to discourage farmers' bush plot activities because they diverted labour from the project. If a holistic view of the farmers' activities had been taken, it would have been realized that the role of bush plots is very significant in the household budget of peasant farmers, providing an extra source of income for minimal capital outlays. The project staff and management failed to investigate the general activities of the farmers outside the irrigation scheme. Hence, it was not possible to synchronize project activities with the other farming activities of the tenant farmers.

Conflicts between farmers' strategies and project objectives

Disharmony frequently developed between the objectives of the project and the strategies which farmers devised to manage a wide range of resources. This resulted in conflicting demands for the allocation of labour and the alienation of farmers from the project, particularly in relation to seasonal labour constraints, the household division of labour, and multiple enterprise strategies.

Seasonal labour constraints

In the Weija area, crop farming, hunting and fishing have their own specific periods in which labour peaks occur. Labour bottlenecks are most felt between the months of August and October (Botchway, 1988). This factor has important implications for the planning of development intervention in the area. The designers of the project's planting program did not take this vital point into consideration. If they had, the planting programmes would have been made flexible enough or relaxed between August and October to allow the project to play a complementary role instead of a competitive one with the other economic activities of the farmers.

In contrast, the project's planting programme for farmers developed from an assessment of optimal weather conditions for planting, and concentrated planting and transplanting activities around September and October, before temperatures and humidity reached levels where fungal problems became more prevalent. Generally, most vegetables planted or transplanted after October in the project area run into production difficulties. These conditions also affect bush plots. Local farmers are fully aware of these weather problems and have developed similar strategic and preventive steps on their bush plots. They start their land preparation for the minor season planting in August and allow transplanting to climax in September. Work on the bush plots therefore competes with work on the project's fields.

The farmers give preference to work on their bush plots during the beginning of the minor rainy season. Two factors influence them:

○ the need for farmers to make sure they have sufficient staple food crops to carry their families through the dry season to the end of the next major season. Clayton (1983) has noted similar strategies among East African cotton

29

producers, where labour priorities are allocated to household food production over the cash crop;

o the timing of planting on rainfed plots is a more crucial constraint than on irrigated plots. Hence, farmers first work on the bush plots, where water is a limiting factor, before they turn their attention to the irrigated plots, where water is not a limiting factor. When scolded for not attending to the work on the project early in the minor season one farmer responded: 'Water on our bush plots can never wait whereas water on the project's fields will never run out'.

During the minor season, farmers arrived late to either abandon weed infested and overgrown seedlings planted on the irrigated plots, or to start new nurseries. Some farmers transplanted overgrown and sometimes flowering seedlings. Farmers were compelled to attempt to transplant these ill-cared-for seedlings because the sowing of new seeds involved additional costs which can be extremely high for some seeds, such as cabbage. Late transplanting and sowing of some crops beyond October (particularly with tomato and watermelon) leads to difficulties in crops surviving the harmattan.

As in other agricultural projects, late transplanting was condemned by the extension staff, without an examination of the myths and rationality surrounding it (Box, 1986). However experiences from Weija have revealed that high yields can be achieved from late transplanting of aubergine. In one instance, 53 day old flowering aubergine plants which were transplanted against the advice of an expatriate staff member gave the highest recorded yield for one of the co-operative units of the tenant farmers. Collinson (1985:72) argues that:

> farmers do not seek technical optima or even optimal economic results from a single activity. They seek to satisfy their priorities through a combination of activities which compete for land, labour and cash; one of which is always scarce, often at the same time in the season. This competition obliges the farmer to compromise the quality of management of a particular farm activity in the interest of the performance of the system as a whole.

Poor quality husbandry is therefore not necessarily bad farm management given the farmers' situation. Such compromises once understood often represent effective leverage points for improvement of farm performance. What needed to be done at Weija was to reorganize the working procedures on the project and to tailor them to the cropping calendar and existing frameworks of farmers.

Alienating the labour of women and children

In the design of the project it was assumed that pooled family labour would form the basis for farm operations. But the project failed to examine implications of the division of labour and the different incentives which influence the strategies of men and women. In the project area, women are traditionally the main organizers of harvesting and they are responsible for marketing. However in the project design, marketing was monopolized by the project's marketing department, and this generally alienated farmers' wives from participating in the irrigated farming.

Within the traditional social system, responsibility for the management of the household economy is divided on a gender basis. Male farmers generally concern themselves with the supply of 'bulk money', for major capital expenditures used in meeting extraordinary expenses such as hiring tractor services and paying for hospital and school fees. Women are responsible for the supply of 'current money', for incidental expenses which arise regularly in the household including the purchase of provisions and food (Collear, 1983). To meet these expenses, women engage in activities which yield dividends in the short term and on a more or less continuous basis (Guyer, 1985). They sell produce like cassava, hot pepper and maize from the bush plots for cash. But there is always a limit to what they can sell because of the subsistence requirements of the family. During the dry season, activities on bush plots drop to a minimum. The women are therefore compelled to diversify their income-generating source and mainly engage in small-scale trading. On the irrigated plots of the project farm, work proceeds uninterrupted during the dry season, but farmers' wives do not consider this as a viable income-generating strategy. One woman commented: 'I will never step on any of my husband's fields on the project because those plots hold nothing good for me and my children'.

The alienation of women from irrigated farming has resulted in a lack of availability of pooled family labour for the project's intensive planting programme. The women and children work solely on their private bush plots and refuse to assist the men in working on the project plots. Children constitute about 50 per cent of the potential labour force per family in the project area (Botchway, 1988). They make significant contributions to most crops. Women's responsibility in feeding children results in their control over child labour, since the provision of lunch can take place around their economic activities either at the market or on-farm. Children thus usually work with their mothers and follow them to the bush plots. If farmers' wives were attracted to work on the irrigated plots, this would also release the labour of their children to work on the project.

If the project had assessed these local farming strategies and perceptions, it would have taken this need of women for 'current money' into consideration, to come to socially acceptable arrangements with the farmers and their households. An agreement could have been reached, where farmers would be allocated a pre-determined percentage of the harvest for their household use. This would have encouraged women to join their husbands on the irrigated plots, since they would have been guaranteed a portion of the harvest for trading. Such a procedure would have blended with the prevailing relations in the local farming system and the women would have made their much needed labour available to their husbands. Similar arrangements elsewhere have proved helpful to both tenants and project authorities (Adams, 1982). Revised harvesting and marketing procedures are the requisites for the participation of household female and child labour on project plots. Changes which undermine women's ability to provide food are undesirable and have ramifications on the whole farming system (Collear, 1983).

31

Multiple enterprise

In addition to their work on the project, male farmers engage in several other jobs like carpentry, fishing and hunting, manifesting the phenomenon of multiple enterprise commonly found among small-scale farmers (Long, 1977). However, the project's work was designed in such a way that farmers needed to work full-time throughout the year with family members if any headway was to be made. In view of the facts that most of the male farmers could not work full-time, and that their wives and children did not accompany them to work on the project's fields, the labour requirements were greater than the actual labour capacities of male farmers. This resulted in several problems associated with different value systems, socio-cultural contexts and procedures, including conflicting demands emanating from, on the one hande, the village farm-economy and strategies involving a mix of part-time enterprises, on the other hand, the commitments to intensive, year-round irrigated vegetable cultivation.

Labour bottlenecks

The project utilized high levels of external inputs including pesticides, fertilizers and mechanized land preparation. High crop yields were needed to justify continuous usage of these inputs, but the cultural practices associated with these inputs required large amounts of labour. Weeding requirements were much higher than under rainfed conditions since weeds also benefit from the constant availability of irrigation water. Each irrigated field on the project needs to be weeded at least three times before the end of harvest. Farmers did not find it easy maintaining any of their fields since most of them came to work on their project fields unaccompanied by their wives and children.

The project's management and staff did not have appropriate concepts of the constraints facing the peasant farmers under the then prevailing conditions. They expected the farmers to make full use of the available technological inputs and the agricultural knowledge transferred to them through the Extension Unit in order to maximize profits. Unfortunately, they did not follow up this expectation with any assessment of the actual farm labour available to individual farmers, to assess their ability to carry out all the necessary agronomic cultural practices and recommendations. If this was investigated it would have been possible for the project authorities to either strike a compromise with farmers' wives by officially allowing them a role in marketing, or to reduce the work load and number of crops cultivated on the planting programme. The labour problems of farmers and their inabilities to cope with all the agronomic cultural recommendations were not considered as problems within the domain of the project's staff. They were left for the farmers to shoulder. Such shortcomings in project design, coupled with the lack of labour, led to the general mismanagement of the individualized fields by the farmers and this resulted in serious crop failures year after year.

Lack of feedback from farmers

During the pilot phase of the project, the Extension Unit performed experiments using fertilizers, fungicides and insecticides to find out the yield potentials of various crops. However, at the implementation stage of the project the research component was phased out. The general crop performance data which were obtained through the initial research were hypothetical. They made no allowance for what to expect on farmers' fields in the event of labour shortages, delays in transplanting, delays in weeding and other anomalies which in reality would be more representative of the life situation on farmers' fields. Under the farmers' management, several crops did not perform as expected. While the farmers called for changes, no research component existed to monitor and evaluate the performance of the cropping system. Without any institutional framework for research, problem identification and analysis, the Extension Unit took a defensive stand in its interactions with the farmers and was not receptive to ideas and information from them. Local knowledge was suppressed and farmers' preferences and evaluations were disregarded, helping to frustrate and choke out the much-needed return flow of information from the field.

The kind of knowledge which was available to the Weija project staff (both local and expatriate teams), was limited to the engineering and agro-biological sciences. This made it very difficult for the project staff to understand the cognitive maps and perceptions of the farmers. With insight into local knowledge, sociological phenomena, group dynamics and the general behaviour and attitudes of rural populations, the project might have had a better grasp of the problems which emerged.

New approaches on the project

Recently, there has been an attempt to reorganize the project's activities and to develop research work which will responds directly to some of the problems posed by the farmers. This new approach takes their farming strategies, conceptual frameworks, and social and cultural factors into consideration for the general development of support services.

Research

Research work is being carried out into late transplanting and its effects on the yield of some hardy plants. If the results confirm observations that a delay in transplanting does not seriously affect yields, it will be possible to fit the project's programme into the local farming system, allowing for late transplanting of some crops on irrigated plots following work on bush plots. Research is also being carried out into the extent to which cow dung introduces weed seeds into plots and enhances the germination of dormant weed seeds already present in the soil. This follows rejection of the use of cow dung by a number of farmers on the basis that it introduces weed seeds onto their fields.

The new approaches being introduced find compromise solutions, which are not only based on technical agronomic considerations, but will enable the project to incorporate a knowledge of farmers' circumstances into the design of technologies (Clayton, 1983:139).

Institutional credit

A programme of institutional credit for farmers also exemplifies a new approach which seeks to root project activities within the fabric of rural life. Shortage of capital is a serious constraint for farmers. While loans have been provided by rural banks for farmers in the past, recovery of loans has been a serious problem. However within the indigenous lending system, based on rotational credit associations (*susu*), there are hardly any problems of credit default.

Farmers perceive government as an impersonal institution which dispenses resources freely, and the rural banks have been unable to develop a system of provision of guarantor or collateral which fits into their way of life. As a result of misconceptions, farmers are unwilling to commit themselves towards honouring repayment of loans. In contrast, *susu* associations revolve around personal relations, and failure to repay debts involve sanctions of ridicule and loss of reputation, which farmers are anxious to avoid.

To devise a credit scheme which will have an impact upon the social life of the rural areas, the project has developed conceptions of collateral which reflect the valuation of farmers rather than of the formal sector. Thus while a mud house may not be expensive, it can be valuable collateral since it also carries social value: the stigma of losing a house will be sufficient pressure to encourage farmers to settle their debts. To further encourage payment, loans are made to teams, allowing the farmers to select one of their members as a guarantor, who pledges his house. The sanction that if the team defaults on a loan one of its members will lose valuable property is a powerful deterrent against default. The strategy also places the guarantor in a supervisory role, ensuring that the members of his team are working hard. At present this form of credit organization is being tried out with 15 farmers.

Implications for development projects

Rural development programmes need to be made to fit into the social systems and strategies of people in rural areas. To achieve this, development workers will have to unearth the indigenous foundations of rural life and build upon the potentials within local productive activities (Korten and Klauss, 1984). Forms of development need to be initiated which encourage the active participation of the farmer.

This paper has drawn attention to areas in which it is necessary for development workers to understand the wider implications of farming and socio-economic strategies. This includes the phenomenon of multiple enterprises, an important indigenous strategy for survival in fragile and uncertain environments, and the gender division of labour. Strategies which do not consider the role of women in the farming system and which introduce changes which undermine women's economic activities may have a serious impact on the whole farming system. The operationalization of fine ideas generated in the formal research sub-system by seasoned academicians is no guarantee for project success. Such fine ideas may yield poor results, not because farmers and rural folks in general are inefficient or uninterested in novel farming techniques and technologies, but because the intervention strategies in question may not have adequately analysed the farming systems operating in the area in order to tailor the vital elements of new technologies to prevailing conditions.

Farmer experimentation and changing fallow ecology in the Krobo district of Ghana

Kojo Sebastian Amanor[*]

The impact of degradation on the farming system in the forest ecotone of southeastern Ghana is examined within the context of farmers' adaptive responses and innovations. Increasing degradation has resulted in unreliable rainfall and infestations of savanna weeds and exotic pan-tropical weeds, which create serious labour problems and disrupt the dynamics of the minimum-tillage, bush fallowing system. Farmers have been forced to revalue their strategies and are beginning to experiment with modes of conserving forest pioneer species and integrating them into agricultural systems. They have rejected the planting of fast-growing exotic species, which have caused problems in their fallows. These rejected strategies are those which are being developed by formal agroforestry. The importance of local knowledge in the definition of problems and prioritization of research, and the implications for the research system are analysed.

The Manya Krobo area of Ghana, situated in the south-east ecotone of Ghana in the area of the Volta Lake, used to be a major oil palm, cocoa and food-producing area. From the 1930s to the 1960s its wholesale food markets were the largest in Ghana, feeding one third of the non-agricultural population. In recent years food production has declined considerably in Krobo, to the extent that in some settlements farmers are now purchasing food on the market to supplement household production. Declining yields are related to problems in the farming system, associated with increasing dessication, deforestation, climatic change, and invasion of farm plots by exotic and savanna weeds. These problems are a result of both local farming techniques and extraneous factors, including the spread of exotic weeds by government services and universities, the formation of the Volta Dam which is likely to have changed the local micro-climate, and possibly global climatic change. Some farmers have responded to these problems by migrating to more forested areas, urban areas and neighbouring countries. Those farmers who remain are increasingly forced to re-evaluate their farming strategies and the potentials of the agro-ecosystem, and to innovate.

---

[*] Kojo Amanor dedicates his paper to 'Michael Kobla Odidja, a farmer at Odometa, Upper Manya Krobo with great environmental knowledge, who taught me how to think systematically about plants and weeds.'
Acknowledgment: The research for this paper was made possible by the support of the United Nations Research Organisation for Social Development (UNRISD).

Innovation and experimentation are concentrated in the more degraded settlements, along the forest edge. Innovations include fishing, cattle rearing, and irrigated floodwater retreat farming on the banks of the Volta Lake. This paper focuses on farmer experimentation with regenerative technologies and their efforts to redefine and adapt techniques of bush-fallowing to present day realities. It examines farmers' responses to problems in their bush-fallowing system, contrasts this with strategies developed in formal agroforestry, and explores the types of linkages between the formal and informal systems which may enhance farmers' experimentation.

Research methods

The aim of field research was to investigate how farming strategies, perceptions and livelihoods varied in relation to changes in the environment. A survey questionnaire was carried out with 178 farmers along a 20km stretch of bush path in which conditions changed from forest bush to grassland, transecting the micro-ecological zones running from the wet deciduous forest interior to the perimeter. Five different localities were sampled along this route, with perceptible changes in environment used as the criteria for selection.

The research approach was flexible, combining both formal and informal methods. A standard set of questions was asked to gather directly comparable statistical data which could be aggregated. However, questions were continuously adjusted in the light of different nuances in changing micro-environments and new insights. This was made possible by the fact that the researcher participated in all the interviews. In addition, an informal exchange was incorporated into the interview, and visits were made to a majority of the farms, in which soil samples were taken and notes made on the nature of the flora on farms. While the survey generated a mass of useful data, farmers' innovatory capacities only emerged in the course of visiting farm plots.

The majority of farmers interviewed were smallholders: 75 per cent of holdings (total farmland at disposal of the household head including fallow) were under six hectares (ha) and 50 per cent under four ha. Women constituted 35 per cent of the sample. Less than one per cent of the farmers purchased agricultural service inputs.

Changing farming strategies

The Krobo farming system has undergone considerable change, adaptation, and experimentation over the centuries (Amanor, 1992). During the 1940s a system of pioneer frontier cultivation associated with cocoa plantations was replaced by intensive bush-fallowing with food crops. The transformation was a response to changing ecology, and to the increasing dessication of the fragile and dry semi-deciduous forest which prevented cocoa from thriving. The decline of the frontier of virgin land and scarcity of land necessitated that a system of pioneer cultivation along a moving frontier be replaced by intensive bush fallowing. Medium fallows of three years were introduced, within clearly delineated farm-strips belonging to individual farmers.

Population pressures on the land have intensified in the last 30 years, particularly since over 20 per cent of Manya Krobo land was inundated in the formation of the Volta Lake in the early 1960s. The main response of farming communities to land shortage has been outmigration rather than resorting to permanent cultivation. The Krobo generally adhere to a two/three year fallow system, and hire land when their land has not sufficiently regenerated. Although some farmers resort to shorter fallow periods and permanent cultivation, this results in a dramatic decline in yields. In certain localities new cropping systems are being developed which seek to dispense with the three year fallow and introduce intensified soil-nutrient recycling.

Soil analysis revealed that the majority of farmers have managed to maintain an equilibrium of over 50 per cent of the inherent fertility of the soil. Most farmers felt their soils were still good 'if it rained', and the major constraints in their farming system were related to unreliable rainfall and invasion of the farm by exotic and savanna weeds. The most significant variations in soil occur with the transformation of land to grassland, where soil fertility drops quite significantly (Amanor, 1992).

Bimodal rainfall results in a major and a minor cropping season within one year. The major and minor 'farms' are usually prepared in different locations. The minor farm (planted in September) is usually prepared on lightly regenerated land, since the farm calendar and short dry season does not allow a long period for clearing land. Maize is often planted as a pure stand in the minor farm, which is then abandoned immediately after harvesting. In major season farms (planted in March and April) the cropping period may extend over a three-year period, but the soil is tilled only once. A system of intercropping ensures that long-duration crops, which can withstand competition with the regenerating bush, are intercropped with short-duration cash crops. This includes:

o maize (short duration), cassava and cocoyam (long duration)
o garden egg, okra, tomato (short duration) and pepper (long duration).

After the third weeding of the farm, following the maturation of the short-duration crops, the long-duration crops are left to stand in regenerating bush, with minimal weeding. By the end of the third year, when the final crops are taken from the old farm, it is overrun by bush, and enters the period of rest as fairly well-established bush. Thus the system of cropping aims to maximize both the crops which can be taken from the land and the period of regeneration by taking both spatial and temporal dimensions of the plot into consideration. Tree crops, such as oil palm, citrus, pawpaw, avocado and coconut are integrated into the fallow and food plot. The Krobo practice a system of minimum-tillage cultivation, which promotes rapid fallow regeneration.

Farmers have also developed forms of fallow management. Particular trees are considered to be good for the soil, and farmers deliberately preserve these species when clearing the fallow. By the time the plot is abandoned to fallow it is covered with a significant number of medium to fast-growing young trees which are reputed to restore soil fertility (Table 1.1). Other plants such as *Mucuna pruriens* (*tsakatsaka*) and *Centrosema pubescens* (*sentrosema*) are recognized as good soil restorants but are disliked because of weeding problems.

Table 1.1: Soil-restoring plant species preserved by Krobo farmers

Scientific name	Krobo name	Comments
Forest trees		
Milicia exelsa	*odum*	Iroko, prime wood
Nesogordonia papaverifera	*bano*	
Pioneer forest trees		
Newbouldia laevis	*nyabatso*	
Baphia pubescens	*tutso*	leguminous
Ficus exasperata	*slabatso*	
Dialium guineense	*mieletso*	leguminous
Trichilia monadelpha	*gbabglablata*	
Albizia adianthifolia	*papa*	leguminous
A.Zygia	*papaku*	leguminous
Holarrhena floribunda	*osese*	
Shrubs		
Ricinus communis	*kumelo*	Castor oil plant
Solanum verbascifolium	*agbafro*	

Problems in the farming system

In recent years, increasingly unreliable rainfall and the spread of exotic pan-tropical weeds, trees and savanna grasses into the forest fallow has resulted in crisis. This particularly affects the forest perimeter areas, which have been transformed into degraded grassland and scrub.

Weeds

Increasing dessication during the dry season creates water stress in trees and results in annual bushfires on the edges of the forest which penetrate into the forest margins annually, killing off trees. Grasses such as *Panicum maximum* (*go*) and *Digitaria ciliaris* (*limman*) which have invaded the fallow are also harbingers of bushfires during the dry season. During the following wet season they colonize the areas of burnt bush, creating the conditions for further penetration of bushfires and expansion of grassland. This process hinders nutrient recycling and the build up of soil organic matter which is annually burned (Nye and Greenland, 1962).

Major weeding problems also arise from the spread of grass. The Krobo machete-based weeding techniques are not designed to cope with grasses, which need to be tilled and hoed. Failure to dig out the roots leads to the farm plot being overrun by grass and to poor yields. On the other hand, if the Krobo till the soil as savanna farmers, the remaining elements of the forest root mat and seed stock will be disturbed and the land transformed into derived savanna.

Problems also result from the spread of pan-tropical weeds (Amanor, 1991). *Chromolaena odorata (acheampong)* is replacing the diverse herbaceous layer of fallow shrubs in the wetter more forested areas. *Chromolaena* was originally introduced into Ghana by the University of Ghana Botanical Gardens in Legon, from where it has spread into the forest. While *Chromolaena* is easy to clear it regenerates rapidly and results in major weeding problems. This causes considerable stress for long-duration crops. Second-year cassava plots are overrun by *Chromolaena* and the tubers fail to grow large and spoil long before their allocated time in the soil. The vigorous growth of *Chromolaena* threatens the principles on which the cropping system are based. An old farm now requires weeding to yield good crops, and farmers have insufficient labour to cater for these new requirements. *Chromolaena* encourages rapid early nutrient cycling. But over a longer period, rates of recovery may taper off and compare unfavourably with a more diversified fallow. *Chromolaena* has received a mixed reception among farmers. Some recognize its good soil-restoring capacities and argue that it has prevented their farms turning into grassland. Others are concerned with its impact on the biodiversity of the fallow and weeding predicaments.

Exotic trees

The introduction of Cassia siamea and *Azadirachta indica*, fast-growing, rapidly-regenerating exotic trees, poses another problem. These trees were widely promoted by the colonial agricultural services and planted by the Krobo as fuelwood trees to replace the forest trees they uprooted in the cocoa era. These hardy fast-growing trees have spread widely and have become dominant trees in some areas, suppressing the regeneration of forest species. In addition to suffering from climatic stress, the fallow seed stock is also under attack from these aggressively competitive 'ecological pollutants' (Tuffuor, 1992). Paradoxically in more forested areas, where these trees are not dominant and not planted, farmers complain of fuelwood scarcity. In degraded scrubland areas these exotic trees have resulted in a plentiful supply of fuelwood both for home consumption and sale. It is not clear to what extent cultivation of these trees has resulted in the degradation of these scrublands. Stress caused by dessication may also have allowed these trees to extend their domain and colonize land occupied by dying forest trees. *Cassia siamea* and *Azadirachta indica* are now disliked by farmers because they are hard to clear and provide poorer rates of regeneration than a diversified fallow. *Cassia siamea* is recognized to have a beneficial effect on maize, but is considered bad for cassava, an important staple in the lean season. In some areas *Leucaena Leucocephala (glauca)* completely dominates the fallow. This plant was originally introduced as a pole and fencing wood but has aggressively taken over the fallow. It is disliked because it is difficult to clear and rapidly reproduces. Land dominated by this plant is often not cultivated.

Farmers' adaptive responses and innovation

The most degraded areas near the Volta lake have become centres of experimentation and innovation. Further into the forest interior, in less degraded areas, farmers are more cautious about experimentation since they are gaining better yields and have more to lose from introducing new unproven methods. It is often presupposed that the farmers most likely to introduce innovatory changes are richer 'progressive' farmers. In this case, the main innovators are poor smallholder farmers and the innovations are in natural resource management. Richer farmers are more concerned with innovations resulting from the adoption of high inputs.

Experimentation is occurring within both the cropping system and modes of managed fallow regeneration in areas around the grasslands. The most successful innovation has been in cowpea cultivation, and its incorporation into a system of crop rotation, which is evolving to replace long-duration intercropping. The major problems influencing adoption of this innovation have been poor rainfall, poor rates of regeneration on grassland, and the long period required to clear grassland for cultivation. On average, 49 days are required for clearing one *kpa ngwa* (0.4-0.6 hectares) of grassland as compared to 13 days for clearing one *kpa ngwa* of forest bush. By introducing this legume farmers aimed to extend the cultivation cycle to three annual crop cycles with variations of rotations of maize, cowpea, and cassava. The cowpea rotation leads to improved soil restoration. Farmer are still experimenting with adapting optimal crop rotations and many variations are being tested. Cowpea has proved to be a good cash earner and has spread widely into more forested areas, where it is replacing minor season maize. Cowpea yields are proving more stable than minor season maize under lower rainfall. However in the more forested *Chromolaena* dominated areas, cowpea yields are lower, and its systematic integration into a system of crop rotation is absent. In these areas the main incentive for adoption is favourable market price.

A second cropping innovation involves replacing long-duration cassava with short six-month maturing cassava. In the more forested areas this enables a crop to be gained before the cassava is overrun by *Chromolaena*. In grasslands this enables systems of extended crop rotation to replace the long duration maize/cassava intercrop.

New systems of managed fallow regeneration are being developed which attempt to encourage improved fallow regeneration in grassland areas. In areas where grassland is beginning to threaten, farmers are developing strategies of preserving all young tree seedlings which germinate. When clearing the fallow they take care to preserve not only young trees but also the young tree seedlings in the undergrowth. It is likely that in the near future this stock of young pioneer frontier trees will become the subject for the innovation of new systems of agroforestry.

New prototype agroforestry systems are emerging. These aim to manage the fallow to maintain a favourable balance of rapid nutrient-cycling biomass and improved soil regeneration. One system aims to preserve high numbers of particular species, such as *Newbouldia laevis*, within the fallow. This tree is particularly suitable for agroforestry, since its slender crown does not shade out light from crops. Root and leaf shoots and seedlings are weeded around. The managed fallow plot is then brought into cultivation without burning or clearing the trees. Where particularly high densities of *Newbouldia laevis* (three metres apart) occur, the trees are pollarded and the cut branches left as a mulch on the ground. All the elements of alley cropping are present, with the exception of planting of neat rows of exotic fast-growing trees. Crops incorporated into the agroforestry system include maize, cowpea, cassava and plantain.

Innovatory agroforestry systems are characterized by the absence of planting of tree species. Farmers claim that they do not plant trees because they don't know what species to plant, and that forest trees are difficult to germinate in nurseries. The latter point is an accurate assessment of conditions of semi-deciduous forest, where seed viability is often low and forms of reproduction through root stolons, shoots and coppice regrowth are of critical importance to regeneration (Janzen, 1975; Ewell, 1980; Nyerges, 1989). Earlier experiences with planting of firewood trees have led to a reevaluation of planting strategies. The newer innovations are based on a discovery of principles of synergism: an attempt to harness the regenerative energies contained within forest succession and adapt farming techniques to enhance these principles. Recent innovatory techniques have developed within an interactive framework with the environment and are concerned with the systems dynamics of environmental regeneration. These innovations have come into being in the last five years and are being experimented with by a few farmers.

Farming knowledge in Upper Manya Krobo is adaptive, interactive and innovatory, rather than a traditional system of stable land management. Farming knowledge arises from careful observation of the natural environment and its reaction to the actions of the farmer. The dynamics of experimentation, innovation and problem diagnosis are intimately connected with the processes of labour. This results in a holistic approach which is concerned with synergism, systems dynamics, and the interrelationship of the energy cycles within the farm environment, rather than the isolation of single characteristics such as yield.

The major constraints for farmers to engage in experimental activity include poverty, the lack of a developed economic and social infrastructure, a depressed artisanal sector, and lack of access to scientific and technical support. Even the most talented of researchers would be frustrated and unproductive working in such marginal conditions.

Constraints within scientific agroforestry

In recent years, formal agroforestry has been given a high profile in Ghana and alley cropping is being promoted as a 'green technology' which will solve the environmental problems of the small farmer. Alley-cropping demonstration farms have been established within all the regions. The technologies have been disseminated from international agricultural centres. However, little data exists on the specific conditions of farming systems in Ghana, the nature of fallowing regimes, farmers' responses to land shortage and degradation, and the physical processes of land degradation in agricultural systems. Technologies for environmental regeneration are being developed and disseminated without any proper diagnosis of the conditions which the technology is supposed to ameliorate. Thus the input of information and technology from international to national centres masks the poor collection of data on national farming systems. National agricultural research is expected to fine-tune technologies generated in international centres to the conditions of specific national farming systems. But without the development of other types of exploratory social and biological research, adaptive research is hampered and becomes largely concerned with the replication of trials devised on experimental plots in international centres.

International agroforestry research is constrained by a 'transfer of technology' approach, which demands that widely adaptable technology is devised for dissemination to diverse national adaptive research centres. While elements of agroforestry have been drawn from farming systems throughout the world, the research process has been driven by the need for homogenized, cosmopolitan technologies, by an inability to test the wide range of plant species and techniques developed by farmers throughout the world, and by an incapacity to engage in formal testing on a scale which can incorporate a wide diversity of environments (Rocheleau, et al., 1989). Formal research is also limited by temporal constraints: researchers need quick results and trees take a long time to grow. Thus agroforestry research has focused on robust, rapid-growing, leguminous trees. But trees which grow rapidly can easily become serious weeds.

Two of the major alley cropping species promoted in Ghana, *Cassia siamea* and *Leucaena leucocephala*, are considered problem weeds by farmers in Upper Manya Krobo. Other studies, specifically on farmers' experiences with alley-cropping, indicate that labour constraints in managing the tree component are a major problem (Gyasi, 1991; Owusu, 1990). Research into agroforestry systems has tended to focus on aspects of yield and soil fertility, without paying enough attention to problems of labour requirements. Farmers in contrast may prefer slower growing trees which are easier to tend. Farmer experiences with planting fast-growing trees in Manya Krobo also point to several problems which may eventually exacerbate degradation. In the light of this experience farmers have moved away from planting exotic species to preserving pioneer species, the main agents which establish the process of forest regeneration.

Agroforestry has largely been conceived of as a commodity discipline concerned with the interaction between a few fast-growing exotic trees and crops. It has failed to integrate the relationship between agricultural systems, agroforestry systems and the larger forest environment into research design. It has failed to analyse the synergetic processes which characterize fallowing systems and the breakdown of these processes under degradation. A major concern of Krobo farmers has been the increasing labour requirements of farming degraded land for diminishing returns. Much of their experimentation is concerned with solving labour problems and finding new energy saving farm management techniques which harness the regenerative energies of the forest fallow. While the Krobo look back to the low energy inputs of farming new forest land, alley cropping only offers a high energy alternative, using plants about which farmers have doubts on the basis of experiences of managing them in their fallows.

Decommoditizing research and building farmer linkages

The commodity-oriented approach of agricultural and agroforestry research results in a failure to examine the processes at work in the larger environment of which agriculture constitutes a subsystem. It results in failure to involve farmers in local-level research and to incorporate the vast wealth of knowledge and experience which farmers have acquired into prioritizing of research. The standardization of commodity-research approaches also hides the lack of knowledge researchers and extension officers have of the species composition and interactions of forests.

New approaches to research are needed to overcome these constraints. These need to go beyond the goals in existing researcher-led programmes of promoting farmer participation for the fine-tuning of research centre technologies so that they are taken up by farmers. The co-option of farmers into on-farm testing of research station technologies may serve to undermine their independent traditions of experimentation (Richards, 1987). The task of fine-tuning technology of general relevance to a wide variety of environments may also be beyond the scope of the formal research system.

An alternative approach is to view the whole process of experimentation as fundamental to agricultural systems, as an essential part of the farmers' craft. Farming communities do not constitute consumers of agricultural technology but resource communities involved in research at the level of the locality (Röling, 1990). Given that much remains unknown in existing scientific knowledge of tropical environments, particularly on the interaction of species within their boundaries, a top-down approach to regenerative technologies is misplaced. An interactive approach imbued with a spirit of discovery and enquiry in collaboration with local people and a recognition of the validity of local environmental knowledge is relevant. Such an approach would enable the totality of knowledge contained within the agricultural system to be brought to bear in solving existing problems. It would enable the adaptive abilities of farmers to be harnessed to promote a continuous process of innovation and adaptation based on the interaction of the farming system with the natural environment. This requires a new orientation in agricultural research, which aims to strengthen the autonomous experimental traditions of farmers, and enables them to make an input in the formulation of problems and possible interventions. The challenge is to develop adaptive research within localities and to build capacities within national research centres to both monitor and support farmers' experimentation. This requires fundamental changes within the present international structures of research.

Farmer experimentation and the cosmovision paradigm

David Millar[*]

The processes involved in farmer experimentation are examined in the light of experiences in Northern Ghana. It is argued that farmers' world outlooks and religious sensitivities are important factors influencing modes and procedures of experimentation. The implications of this for frameworks of research are examined. Suggestions are made for new research approaches which redefine the relationship between farmers, extensionists and researchers and place the agricultural knowledge and information system within the sphere of human cultural and value systems.

Official agricultural research and extension approaches have had problems dealing with factors which do not appear to have anything to do with agronomic variables, but which influence the modes through which farmers carry out agriculture. These factors are often rooted in the cultural and spiritual world outlooks of farmers. This paper examines the importance of such world outlooks or *cosmovisions* for agriculture, and the ways in which agricultural research and extension can benefit from an understanding of farmers' life worlds (Long, 1990). It examines the forces behind the decisions and actions of farmers in the process of experimentation, analyses the processes involved in farmer experimentation, and shows that these have significant parallels with scientific procedures and can be classified both according to scientific logic (Latour, 1983) and the logic informed by their cosmovisions (Haverkort *et al.*, 1992). The paper suggests a new framework for research and extension which can integrate these concerns.

The cosmovision paradigm

The concept of cosmovision originates with PRATEC, a Peruvian NGO, and was developed in the context of recovering the indigenous culture and agriculture of Peru as the basis for improving the lives of the majority of people. Agriculture is conceptualized as the product of a unique indigenous culture in which technology is embedded in nature and of a holistic world outlook based on a dynamic framework of continuous interaction with the environment. People are

---

[*] David Millar dedicates his paper to Farme Dachil from Yachedo near Tatale in the eastern corner of the Northern Region of Ghana. 'Not only for the rich ideas he gave me, but also for the greater and richer portion which he did not share with me.'

conceived as part of nature and all parts of the environment are interrelated (PRATEC, 1991).

This paradigm can also be used in understanding the integrated relationship between spirituality, nature and agriculture (material culture) among small farmers in Northern Ghana. This integrated relationship is reflected in the position of the soothsayer who combines a role as intermediary with the spiritual world alongside functions in health and agriculture. For the small farmer, the spiritual dimensions are consequent upon worldly actions and are reflected in the treatment of nature in the processes of gaining a livelihood and cultivating food. Spiritual manifestations impact on the relationship between people and nature and on the technique of agriculture. Thus one farmer in the Tallensi area, commenting on the possible causes of decreasing soil fertility, saw collaboration with other lineages and religious functionaries as vital to problem-solving and diagnosis:

It is not my problem alone. My wife told me her father and several other farmers in that area who cultivate around the same piece of land have the same problem while others of another section have very good yields without any fertilization. To find a solution, the head of their lineage would have to consult with the elders and the chief, after which they would have to consult the soothsayer, ask the earth priest to make a sacrifice to the ancestors, especially the founder of the village, and to send their petitions through the gods to the Almighty God. While these consultations are going on, since it was the Almighty God who has given us the land, vegetation, water and livestock, some of which are for the use of the gods only, we would keep adjusting our resources, methods and even add some fertilization from animal waste to see whether things would improve. I am sure both 'medications' are necessary. Even with the Muslim or Christian God, the final outcome when people are sick is a result of the combination of prayer and medicine.

The institutions which characterize the social, political and religious worlds, the chiefs, soothsayers and earth priests (*tendana*), mediate between the supernatural and natural worlds, communicating messages through a pantheon of ancestors, lesser gods and the almighty God. These messages have a bearing on the soil, wind, rain, wellbeing of the people and the performance of crops and livestock (Figure 1.1).

The issues of rewards and punishment, as manifest in harvests and health, are related to how humans treat nature from the perspectives of the spiritual world. The acceptance of innovation by communities is very much guided by consultation with the ancestral spirits and the responses of the natural world after such consultations. As one farmer commented:

The white-man came to us and told us that he wanted to plant trees in our village because trees are of great benefit to the soil. We told him to let us consult our elders, our ancestors and our gods first. He allowed us. So we did our consultations and the gods showed us where to plant and where not to plant the trees. We communicated this to him but he refused and said

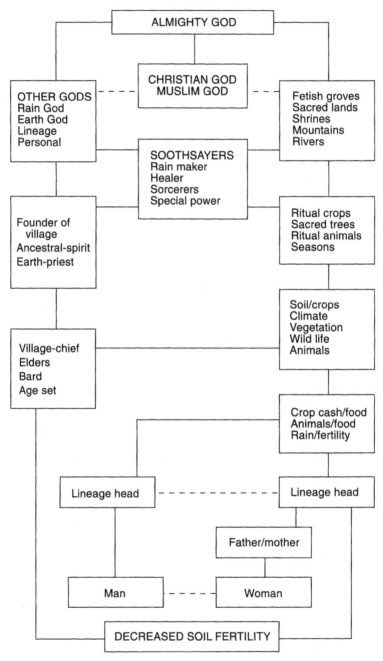

Figure 1.1: *Cosmovisions of rural people, Northern Ghana*
Source: Millar, 1992

he wanted to plant them everywhere in the village. So we collected some and planted them where the gods have shown us and refused to help him with where he wanted to plant the others. We planted ours here and they are still standing but all of his were planted over there. Do you see any trees standing there? They are all dead. The gods were angry so they destroyed them all.

Farmer experimentation

The religious world also influences processes of farmer experimentation and creates impetus for particular types of experimentation. In addition to curiosity, problem-solving and adaptation experiments (Rhoades and Bebbington, 1991) farmers in Northern Ghana also engage in social or peer pressure experimentation. Farmers' cosmovisions are a powerful driving force for such experimentation, in which cultural and religious factors are important. This type of experimentation requires that certain varieties must be planted each year despite undesirable production qualities, and that each year they must be planted differently from the previous year, as indicated by the soothsayer. This gives rise to a paradigm of continuous experimentation. Farmers monitor their neighbours' experiments, and those who repeat the same experiments are considered to be poor farmers who will be sanctioned with poor harvests by the ancestors. Such beliefs encourage farmers to consolidate their skills and planting materials.

Processes of peer pressure experimentation

As with formal agricultural technology development, farmer experimentation follows logical processes. However, they involve distinct social units in different phases. The stages of experimentation include:

Problem identification. This involves family members. Problems are initially identified by the husband and wife team and then discussed with the head of the household and other colleagues. The problems include old recurring issues, recent developments resulting from previous problem-solving activities, and new unforseen predicaments arising in production.

Testable hypothesis formulation. Before the onset of the farming season households compare the present problems they have identified against their past experiences, and formulate hypothesis which will guide their experimentation.

Design. This includes consultation with the family and soothsayers and determines the siting and lay out of the experimental plot, the factors to observe and the indicators to be measured for determining success or failure. Since problems have been discussed with other farmers from the problem identification phase, farmers are aware of the experiments being carried out by others and make inputs into their design. As a result farmers integrate their experimentation with those of their neighbours and utilize the results of others in evaluation. This results in a process of experimentation which has the features of replicated multi-location trials under different management regimes. Experiences are shared during the cropping season and at the end of the season detailed discussions are held under the shade of trees, resulting in a thorough evaluation of experimentation.

Testing. This merges into the entire farm operation and the main aims are to find if the experiments fit into the cropping calendar and labour profile. Mental records are kept charting the course of the experiments. With the testing of new varieties, experimentation is often systematic and closely monitored for evaluation. Other types of experiment reflect an 'adaptive rationality' which may appear to be based on trial and error, to be non-systematic and chaotic.

Validation. This process starts from the formulation of the hypothesis. Experiments are validated under different parameters including social-cultural factors. These parameters may include the quantity of seed required for sowing, labour required for cultivation, harvesting and transportation, the ability of the crop to provide an all year round source of food, quantity of harvest, taste of crop and sensitivity to the growing environment. Experimentation is also interactive and iterative and in the course of testing, the hypothesis may change. Hypothesis development does not follow the strictures associated with formal science.

Evaluation and utilization of results. Analysis and utilization of results are simultaneous processes which start very early in experimentation, unlike in formal science where they occur at the end of experimentation. Small farmer experimentation is highly flexible and in the process of testing farmers may begin to process and utilize results, observing neighbours' experiments and using some aspects immediately. Nevertheless, an end of season evaluation of the outcome of experimentation is essential. This is carried out in different social units, including the conjugal unit and with other members of the household, with neighbours or friends who have integrated their experimentation through joint consultation, and with other members of the village in conversation at the end of the season. At the beginning of the new farming season, farmers often discuss and analyse the results of the previous year's experiments in order to map out the uses of the results.

The inherent differences encountered in the processes of experimentation between science and farmers are significant for the processes of technology development. Recognition should be given to functional and institutional overlaps and the fact that the processes of testing and validation and dissemination of findings, results and experimental design are all carried out simultaneously.

Incorporating farmers' cosmovisions in research

Farmers are simultaneously researchers, extensionists and users of technology, and they rely on their cosmovisions to direct the course of their actions. There is a need to develop alternative methodologies for agricultural technology development which integrate the extension sub-system into the research sub-system and which take cognisance of farmers' worldviews. This could be done through allowing farmers to process technology through their ancestral worship, or through developing technologies *in situ* with them, to develop a synergic interaction between both traditions of research. The performance of extension and research would be enhanced by the use of a dialogiçal approach which enables farmers, extensionists and researchers to analyse problems jointly within the farmers' environment, and which takes their cosmovisions into consideration.

Extension staff need to be reoriented towards participating in farmers' programmes and understanding the importance of rural people's knowledge (RPK) and cosmovisions. Experimenting farmers can be resources for agricultural technology development, with researchers and extensionists as catalysts. This would facilitate the exchange of information, enable farmers to understand the processes and content of formal science, and help researchers to gain a better understanding of the research procedures of farmers. This would promote improved interaction and cross-fertilization of ideas between professionals and farmers, enabling more flexible technology and extension messages to emerge. Tools for the integration of the sub-systems of farmers, researchers, and extensionists are important, as the maintenance of distinct professional domains hampers participatory technology development.

Implications for the research system

In agricultural technology development practice, formal science is more concerned with the generation of *products* rather than *sustainable processes* and

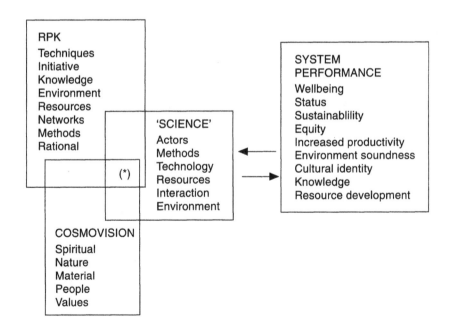

Figure 1.2: *Synergic integration of RPK, science and cosmovision*

Source: Millar 1992

extension views farms as *receptacles* for technology rather than *learning systems*. Continuous dynamic processes are broken up or regarded as static. RPK can be used to further process and transform inputs from formal science, acting as a strong user-driven configuration to enhance synergism.

A synergic agricultural knowledge and information system (AKIS) demands an integration of cosmovisions, indigenous agricultural knowledge and formal research and extension, resulting in processes and products which are greater than the sum of the individual parts (Röling, 1990). This is reflected in improved status of the farmer, increased productivity, stability, sustainability, participation and equity, socio-cultural development, and the total resource development of the actors. The challenge is to integrate the AKIS with the human activity system (HAS) (Checkland, 1985), from which it has been abstracted. The desired integration requires the management of the interfaces (Long, 1990) between RPK, science and cosmovision (as identified in Figure 1.2). The starting point is to pay attention to the cosmovisions of farmers as the bridge for the development of a new type of training and dialogical development.

Selection strategies of rice farmers in Sierra Leone

Catherine Longley and Paul Richards*

This paper examines the cultural and socio-economic factors underlying farming strategies, seed selection, and farmer experimentation drawing on case studies from the Mende and Susu areas of Sierra Leone. The paper draws attention to the social embedding of indigenous technical knowledge and argues that this knowledge has to be understood within its social and cultural context, before it can be decoded and incorporated into appropriate conservation and development programmes.

Recent interest in the use and conservation of landraces serves to highlight the need to understand the social and cultural environments in which crop varieties are maintained. Human selection is regarded as a most important feature in defining landraces, yet both the criteria and methods used by farmers in their selection processes remain largely unknown (Hodgkin and Ramanatha Rao, 1992). Natural selection and environmental adaptation also contribute to the genetic integrity of landraces. In a discussion of crop evolution, Simmonds (1979:11) states that 'the act of cultivation is perhaps the peasant's most potent contribution' to natural selection, implying that the human factors in the overall development and maintenance of crop genetic diversity are perhaps rather more complex and far-reaching than might be assumed.

This paper draws attention to the social embedding of indigenous technical knowledge concerning rice germ-plasm, describing the processes through which farmers select and conserve distinct classes of rice in two different communities in Sierra Leone. In this case, indigenous technical knowledge is generated by social practices involving strategies of labour management and gender-based division of labour. Furthermore, discovery of new technical knowledge, or adoption and rejection of technical innovations, takes place within moral orders that are specific to given societies. This is why it is often difficult to provide a reliable 'decoding' of the significance of 'local knowledge' in scientific terms without careful prior analysis of the social and cultural context. In essence, we need to understand not only what farmers do but why they do it and how they understand what they know and do (Thrupp, 1991).

---

* Catherine Longley and Paul Richards dedicate their paper to Chief Alimanmy M'fansomani Sill of Sabuya and Foday 'Capitalist' Dumbuya of Kukuna.

51

Evolution and social significance of rice types: a Mende case-study

Mende upland rice farmers in Sierra Leone do not farm fields, they manage the complex cycle of vegetation changes associated with forest regeneration in systems of bush fallowing. The major historical limit on upland Mende rice cultivation has been that of restricted labour for forest clearance and land shaping. The earliest farm sites perhaps were clearings caused by natural tree falls, but enlarged in swampy areas especially by the action of elephants and other forest 'bulldozer herbivores' (Kortlandt, 1984). In a number of cases the elephant is also thought to have contributed to the origins of agriculture by excreting undigested grains of rice grazed in distant farms (*helekpoi* 'elephant dung' is a category name for a range of rice types in Mende). The elephant, then, is seen as an instinctive agriculturalist bulldozing vegetation and introducing a diversity of new seed types. Early Mende farmers, reflecting upon this example (Hill, 1984), substituted ingenuity for bulldozer strength by experimenting with rice germ-plasm diversity. There are reliable historical accounts from the seventeenth century, when the forests of eastern and southern Sierra Leone were still being settled, of farmers planting up and down slope, carefully matching rice varieties of different duration to soil moisture conditions at different points on the slope as the rainy season advanced (e.g. Dapper, 1668, quoted in Jones, 1983).

Thus, farming strategies focus on developing skills to match environmental conditions (exploitation of natural variations in soil moisture conditions along the soil catena) rather than investments in technologies to control the environment (land shaping and water control). Since not all land is ready for planting at the same time, these techniques help minimize labour bottlenecks at planting and harvest, and maximise the period during which rice is available in the field. Short-duration rices ripen in 90 to 120 days planted on moisture-retentive soils at the foot of valley slopes, or in marshy depressions (*bului*) of the kind once favoured by elephants, serve to reduce the period of pre-harvest hunger. Medium-duration higher-yielding varieties with strong root systems are planted on free-draining upland soils when the rains are well set (these continue today to supply the bulk of the crop). Long-duration types, ripening in five to six months and adapted to variable flood levels (including floating types) are planted in the beds of water courses and in inland valley swamps during breaks between the main upland operations, and are harvested at leisure after the main crop of dryland rice has ripened. Traditionally, Mende farmers do not irrigate: they match their planting strategies to the changing availability of soil moisture on the soil catena throughout the year. They follow the flood.

In effect, then, in labour-constrained circumstances, Mende rice farmers discovered that it makes more sense to develop the 'software' possibilities (manipulate germ-plasm) than to rebuild the 'hardware' of agriculture (level fields and reshape the land for water control, etc.). How was this 'software' approach first worked out? Rice is largely a self-pollinating crop, subject only to small amounts of accidental out-crossing. The simplest method of harvesting, breaking off panicles one-by-one by hand, gives farmers the option to reject off-types as they harvest. Avoiding off-types in the course of panicle harvesting is equivalent to the procedure plant breeders term 'mass selection'. Panicle harvesting stabilizes the main seed types and also brings about a systematic grouping among off-types: early-ripening types will be rogued as they ripen by farmers anxious to secure a little extra consumption in the hungry-season, and longer-duration types will be left in the field to the gleaners. In this way, over a long period of time, panicle selection has resulted in the differentiation of Mende rice germplasm into three distinct duration classes.

Today, social differentiation is evident in the ownership and control of rices belonging to each of these distinct duration classes. There is, for example, a well-documented association between household dependents (women and young people) and long-duration flood-tolerant rice types throughout the rice-zone of the West African coast (Watts and Carney, 1990). In Mende villages, it is common to find, in addition to the large household rice farm, an associated set of smaller private rice plots cultivated by young men and older women especially. Typically, these plots are located in valley-bottom wetlands adjacent to the main farm, since the work cannot be started until the main tasks on the family farm are well in hand. The work has to be slotted in during slack periods on the main farm. Finding seed to plant such a farm is not always easy. Upland gleanings are a potential source. Gleaned varieties tend to be longer-duration types which have been rejected because they were not yet ripe. Cultivation in valley-bottom wetlands acts as a further screening process: casual farming permits no water control, and only those varieties capable of withstanding a variable flood will survive. Hence dependent's rices tend to be of the long-duration flood-tolerant type. The Mende word for these kinds of rices is *yaka*, a word sometimes glossed as meaning 'charity'.

There is a definite prejudice against consuming these long-duration flood-tolerant rices. Mende villagers reckon them flavourless and lacking in nourishment, even though some recently-introduced types are in fact of high quality. Coming last in the harvest sequence, *yaka* rices are not vital to subsistence, and so have a low status in the village 'moral economy'. They are the rices that people *can afford to sell* because they are not needed to build up the community. It is tempting to surmise that they are considered to have 'little flavour' and at times are suspected of causing sickness because they do little to sustain social values.

By contrast, to acquire short-duration seed types and share them among family and friends assumes central significance in repairing the social fabric damaged by climatic irregularity and other misfortunes (Richards, 1986; 1990). Farmers search for short-duration types that ripen ahead of the main harvest because they consider this offers a durable answer to the social evil of indebtedness caused by pre-harvest hunger. Suitable planting material is seized upon, segregated, and regularly tested in small trial plots (*hungoo* in Mende). Farmers are explicit about the need to maintain the pool of germplasm biodiversity for rice through such experimentation and exchange. Informants have told me 'it is the nature of rice, and circumstances, to change'. No farmer believes in an ideal variety to which he or she will remain committed for life. The generations of rice are like the generations of humans: no child is an exact copy of either parent. Rice changes, however, within a framework governed by ancestral forces, the guardians of the moral order. It is through ancestral blessing that familiar and much loved forms reassert themselves from time to time. In Mende this is given explicit recognition in the rice category name *mbeimbeihun* (literally 'rice-within-rice').

Segregation of short-duration types must have been less easy to achieve than isolation of the long-duration flood-adapted types since the principal use of short-duration types is as hunger-breakers. Households short of food are less likely to be in a position to save any such seed to plant the following year. The selection and maintenance of stocks of short-duration rice in order to exploit specialized wetland niches (valley hollows, *bului*, and moisture-retentive river-terrace soils) may thus have depended on a well-developed sense of social responsibility among larger, wealthier households better placed to retain stocks of these seeds. The following historical case could be interpreted as an illustration of that sense of social responsibility at work. In the 1780s, Henry Smeathman, an English merchant-reformer resident in Sherbro country (now part of the Southern Province of Sierra Leone), thought he might greatly improve the efficiency of local rice harvesting by introducing the sickle. Sherbro farmers were most reluctant to accept his innovation, fearing they would be accused of witchcraft. To Smeathman, this obduracy was an instance of the untutored African mind at its most obstructive. As we have seen, however, the viability of the local farming system depended on panicle harvesting to maintain the careful separation of rice types suited to the different soil types of the upland-wetland soil catena. To be accused of witchcraft is (in effect) to be charged with anti-social behaviour. Smeathman's informants might well have feared such an outcome had they adopted his recommendation, since indiscriminate sickle harvesting would have put at risk the coherence of the entire local programme for germplasm management. Smeathman was, in effect, enticing farmers to put private efficiency before the public good. This is an early instance of the dangers of failing to spot the way in which indigenous technical knowledge and social order are intertwined.

Rice knowledge among the Susu of Kukuna: one system or many?

An on-going study of local knowledge systems in rice agriculture in the Kambia District of north-west Sierra Leone has revealed a striking degree of variation in cultivation strategies and attitudes towards innovations among the farmers interviewed. In addition to the considerable differences in cultivation strategies in separate villages, significant contrasts in farming techniques existed among farmers within the same village. This is best illustrated by the case of Kujuna, a Susu town in Bramaia chiefdom (Kambia District).

The Susu are an agricultural people of north-western Sierra Leone and southern Guinea whose main staple is rice. Thayer (1983) describes the Susu as pious Muslims who draw unusually strong distinctions between categories of nature and culture: the natural world is regarded disdainfully as a profane and hostile environment, and only the realm of human culture is considered to be the proper realm of Islam. He states that Susu Islamic ideology systematically denigrates man's sense of union or identity with nature. But do such distinctions exist at all levels of the society, for example, among women as well as men, or among the rich as well as the poor?

The large and newly-built mosque in Kukuna bears witness not only to the central role of Islam within Susu society but also to the relative wealth of the town. Situated on the Sierra Leone-Guinea border, Kukuna is an important centre for the sale of groundnuts to traders in Freetown. A number of local farmers have earned substantial profits from the groundnut trade and are much respected for their wealth. It is these wealthy, senior men who have also adopted interesting agricultural innovations not widely known about in Sierra Leone.

Like the Mende, the Susu make use of natural variations in soil moisture conditions along the soil catena, and farmers have a detailed agro-ecological knowledge, carefully matching rice varieties to the particular niches where they were known to grow best. A popular *Oryza glaberrima* variety called *disi kono*, for example, is an especially resilient and drought-tolerant type, capable of withstanding competition from weeds. For these reasons, farmers often plant it at the top of the farm catena where the soil is typically gravelly and dry. Another farmers' variety, *samban konko*, by contrast, is a much higher-yielding *O. sativa* variety which tends to be planted in low-lying, moist soils where it can grow to as high as 160 cm. The farmers who plant this variety are aware that if it is planted in soils with a lower water content it will not grow to such a great height.

All this might be seen as 'common knowledge', and anyone familiar with these varieties also knows that the grain of *disi kono* remains slightly hard when cooked, but that it is pleasantly filling when eaten. *Samban konko*, on the other hand, is reckoned to be both better-tasting and softer when cooked, though not as filling. But what was not common knowledge to the majority of farmers interviewed is that the seeds of these two varieties can be mixed and sown together on level ground, in the middle of the soil catena, where they will both grow to the same height and mature at the same time. This is a technique practised by only a few farmers.

One farmer, a village chief, described how he had adopted this technique some five years previously, after a drought caused *samban konko* to produce black, empty grains. Since that time, he has been mixing the two seed types and has not experienced any problems. Once mature, the rice is harvested in bunches using a sickle (as opposed to the Mende method of panicle harvesting described above). The mixture is then threshed and milled and cooked in the normal way, and the qualities of the two rices complement each other when eaten. The *sativa* variety provides high yields and the *glaberrima* provides nutritional bulk. This example of local knowledge is indeed impressive in terms of the farmers' detailed knowledge of two rice varieties and their comparative qualities related to drought-tolerance, duration, height, yield, taste, texture and nutritional bulk.

This technique was practised by only a very small number of successful 'big men' in and around the Susu town of Kukuna. Other farmers in Kambia District were apparently unaware of the practice, despite their familiarity with each of the rice types described. Moreover, the majority of farmers interviewed in Kambia District regarded with horror the notion of planting mixed seed types. Rice bought from the marketplace often contains randomly mixed seeds of different varieties, and this is a common cause for complaint among women because mixed rice is more problematic to prepare and cook. In central Sierra Leone, Mende rice farmers regard the presence of *O.glaberrima* types in a field of *O.sativa* as a social disgrace since it is a sign that the farmer ran short of rice and had to borrow seed at planting time (Richards, 1992).

Why has this technique of mixing *O.sativa* and *O.glaberrima* varieties been adopted by only a few male farmers in and around Kukuna, and wealthy, successful ones at that? As a major cash crop in Kukuna, groundnuts are grown in the second year of the crop rotation, after one year of rice cultivation. The labour requirement for clearing the land in preparation for groundnuts is substantial, and only wealthy farmers have the means to secure the inputs of labour necessary to clear a sizeable site. In addition, the site must be sufficiently dry to prevent the groundnuts from rotting in the soil. But the flat, rain-fed sites preferred for groundnut farming are not ideal for the production of the high rice yields which are needed to meet the subsistence needs of a wealthy man's household and his many dependents. Although this risk may be worth it for the potential cash income of groundnuts, the potential for disaster in terms of subsistence rice yields is substantial (Nyerges, 1987). Some of this risk can be averted by planting two complimentary varieties: if the high yields of *samban konko* should fail due to insufficient rainfall, all is not lost since *disi kono* (a hardy variety capable of withstanding drought) will always yield something. In a good year, on the other hand, *samban konko* will provide a large quantity of a particularly tasty and well-liked rice, and this will be enhanced by the nutritional satisfaction provided by the 'bulkiness' of *disi kono*.

This example illustrates that indigenous technical knowledge is not a single, coherent body of knowledge shared by all members of the community. Some discoveries are treated as private possessions and closely guarded secrets. Such secrets certainly exist among Susu farmers: on several occasions interviewees bluntly refused to answer specific questions because the information was considered to be confidential. Could it be that such secrets are defined in part by the need for wealthy Susu men to maintain an air of modesty concerning their knowledge of the natural world? Thayer (1983) comments that the Susu regard *kheme mokhi* (adult men) to be the apex of cultured humanity and that wealth is supposed to enhance a man's moral capacity. Wealthy men are thus greatly admired for their moral respectability as well as for their money. So perhaps wealthy Susu men are reluctant to reveal their detailed and well-informed knowledge of the natural world in order to maintain their status as good Muslims.

Conclusion

The human element in the creation and maintenance of rice varieties, whether this is the result of conscious selection by farmers or the unintended consequence of biological adaptations to particular cultivation practices, is a significant factor in crop development. In the Mende case, panicle harvesting has allowed farmers to select rices for characteristics of both short and long duration. Around Kukuna, it is possible that the farmers' technique of intra-specific rice cropping may increase the potential for accidental crosses and gene flow across sterility barriers, giving rise to the emergence of intermediate rice types.

But in order to fully understand these processes it is necessary to examine the socio-cultural context in which local agricultural knowledge is generated. Expertize in the management of natural conditions, together with a well-developed system of labour organization has, over time, allowed for Mende rice types to be segregated into distinct duration classes, and these are effectively maintained by a set of social norms and obligations at work within the society. The introduction of the sickle to Sherbro farmers, for example, was clearly not in tune with their sense of social responsibility.

Yet innovations which are seen in one context as threatening and 'unthinkable', in another situation might be neutral and 'safe', to be experimented with at will. Among the Susu, different attitudes to nature and agricultural experimentation may reflect an individual's place in society. At a regional level, the practice of mixing two rice species has been mentioned for other parts of West Africa (Merrick, 1991; Clawson, 1985), yet it appears to remain relatively unknown in Sierra Leone, and even discouraged among the Mende. Attitudes to innovation can thus be seen to articulate with social, cultural and economic differences among farmers, both within and between societies.

The examples of local rice management presented in this paper can be used to show different ways in which local knowledge might be examined and interpreted. At one level, local knowledge can be seen to provide western science with 'new' phenomena to be tested, investigated and understood for scientific purposes. The case of mixing rice varieties in Kukuna, for example, may encourage plant scientists to study the effects of intra-specific cropping on the efficiency of water uptake and on the risk of disease and pest attack. While such an analysis may contribute to advances in the fields of plant physiology and pathology, this is unlikely to reveal much about the significance of intra-specific rice cropping at the local level. We argue, therefore, that it is also necessary to study the relevant cultural and socio-economic factors involved. Social and moral values may, in some circumstances, reflect a well-informed understanding of local ecological processes and the related effects of specific cultivation practices (Richards, 1993). An awareness of these values is important in helping to set the standards and objectives for appropriate agricultural development programmes, particularly if formal research into local knowledge systems is ultimately to be useful to resource-poor farmers as well as to scientists.

PART TWO

DEVELOPING LOCAL CROPS

Introduction

Kojo Amanor

A number of researchers working with plant genetic resources have become increasingly critical of existing plant breeding strategies. They are searching for new paradigms and new methods of working with farmers and plant genetic resources. This section reflects new perspectives in the conservation and development of landraces or farmers' crops. It includes descriptions of farmers' crop selection activities, theoretical investigations into the rationale of local crop selection strategies, and the search for new institutional forms which promote collaboration between researchers and farmers in the conservation and development of local genetic material.

Hardon and de Boef and Berg are concerned with developing strategies and appropriate research methods for the conservation and development of local crops. Mekbib and Worede examine approaches which aim to generate more appropriate genetic materials for marginal areas in Ethiopia, through the conservation and development of landraces. Mushita is more concerned with socio-cultural and conservation perspectives, which aim to preserve landraces and the associated cropping system knowledge in Zimbabwe from the onslaught of modern varieties and the increasing commercialization of agricultural research. Van Oosterhout is also attentive to the socio-cultural implication of the different criteria plant breeders and farmers use in selecting and evaluating species and its implication for the generation of technology, the preservation of biodiversity and the organization of research.

Hardon and de Boef, and Berg, critically examine constraints within modern plant breeding strategies in the context of marginal and diverse environments. They argue that two systems of crop improvement exist, with different objectives and strategies: a formal institutional system focusing on maximizing yield and standardizing varieties, and an informal community-based system emphasizing yield stability, risk minimization and diversity within and between landraces.

Hardon and de Boef contend that while the two systems have different objectives they are complementary, with relevance to different situations. Modern plant breeding strategies are constrained by several factors in carrying out research into minor crops and in the adaptation of major crops to ecologically marginal and diverse regions. In examining the objectives of different conservation strategies, they argue that local crop development has an important role to play in that it creates dynamic conditions for the further adaptation of varieties to local conditions. But local knowledge is also constrained by problems of increasing yields while conserving the natural resource base. Local crop development can benefit from methods of modern plant breeding, given that the comparative advantage of local crop breeding is recognized and support structures developed to strengthen farmers' crop improvement activities.

Berg is more concerned with the breeding activities of farmers rather than the conservation of landraces. He examines the rationale behind the plant breeding activities of farmers and argues that this could be strengthened by more sophisticated methods of selection. The main factors which has influenced the divergence of strategies have been commercial market demands on the formal plant breeding sector. Farmers' plant breeding activities can be supported by the provision of enhanced germ plasm from scientific institutions allowing for a broader base for local selection. Drawing on examples from the Sudan and Ethiopia, he argues that local plant breeding activities and community organizations can form the basis of programmes which aim to increase the efficiency and quality of traditional seed development activities.

Mekbib and Worede describe the activities of the Plant Genetic Resource Centre in Ethiopia, which has come to recognize that farmers have made important inputs into the formal plant breeding sector in Ethiopia, and is now developing a number of programmes in collaboration with them. These involve farmers in selecting, evaluating and multiplying materials collected from surrounding areas, or in experimenting with enhanced germ-plasm. By collecting and storing local landraces and making them accessible to the communities, the support activities of the PGRC/E encourages farmers to experiment, knowing that particular landraces can be retrieved from the genebank if the experimental process goes wrong.

Mushita describes the rationale behind the development of an ENDA-Zimbabwe programme to strengthen local small grain selection activities. Farmers seed selection has been discouraged by the formal sector and by agricultural and political policy frameworks which have eroded the basis of traditional agriculture and encouraged the uptake of modern varieties. This has resulted in increasing land degradation problems and insecurity of yields in the marginal communal areas. In some areas farmers have continued to preserve small grain cultivation and indigenous methods of cultivation, and have managed to stabilize yields under conditions of stress. These areas have become the subject of an ENDA-Zimbabwe project which seeks to understand the rationale of the indigenous cropping system and the selection of small grains within it, as the basis for developing local crop conservation activities.

Van Oosterhout examines the criteria which farmers use in selecting sorghum varieties in Zimbabwe, as a means to understanding factors which are appropriate to the conditions under which small-scale farmers operate, and which should be built into the priorities and recommendations of the formal sector. In contrast formal plant breeding, farmers are concerned with a wide range of factors including gastronomic qualities, maturity, agronomic criteria, stability and reliability of yield, and the interaction of crops with other elements within a diversified farm environment. High-yield characteristics, the major concern of the formal plant breeding sector, are not by themselves a dominant criteria. The high-yield package solutions of the formal sector are found to be wanting, and participatory methods, which develop recommendations on the basis of farmers' needs and perceptions, are advocated.

The peculiarities of the development of local crops and linkages between plant breeders and farmers, have implications which transform participatory research methods. In place of the emphasis on building feedback mechanisms from farm to research centre, and developing collaboration in the on-farm testing of research centre technology for fine tuning, the major emphasis is on biodiversity and natural resources, on the genetic materials and knowledge of farmers, and on utilizing, strengthening, monitoring and evaluating the independent experimental traditions of agricultural communities. The issues which are raised have deep socio-economic ramifications concerned with the interrelationship between environment, farming strategy and farmers' perceptions; and the problems of transforming top-down procedures, objectives concerned with standardization, and institutional arrangements which have alienated farmers from both the production of varieties and the determination of the relevant criteria they should meet.

Linking farmers and breeders in local crop development

Jaap J. Hardon and Walter S. de Boef*

Local crop development and plant breeding are complementary activities and co-operation between the two systems is essential and of mutual benefit for plant breeders and farmers. This paper focuses on the limitations of plant breeding in marginal environments and the potential role modern plant breeding can play in the improvement of local crop development, while maintaining the integrity of informal farmer experimentation. It is argued that modern plant breeding needs to investigate the techniques of local crop development and develop a participatory framework for research which will enable new information to be generated to illuminate the role of small farmers in genetic resource conservation.

Since the dawn of agriculture, farmers have played the major role in harnessing genetic diversity by domesticating wild species and adapting crops to a variety of new environments far beyond their original range of natural distribution. When, early this century, plant breeding developed as an applied science of genetics, plant breeders inherited a multitude of landraces which had co-evolved through processes of natural and human selection. This process of co-evolution, in association with the maintenance of a degree of diversity (Hanelt, 1986) in farming systems adapted to local environmental conditions, has provided harvest security and yield stability. A remarkable similarity exists between these cropping systems and strategies employed by wild species in natural ecosystems (Boster, 1985)

In contrast, the development of modern plant breeding has been associated with technical change in agriculture and the rise of the fertilizer and pesticide industries. Plant breeders have focused on raising the genetic yield potential of major crops, by increasing the amount of dry matter diverted to harvested product and minimising the increase of total biomass production. The expression of a higher yield potential in modern varieties is generally based on better utilization of additional external inputs, notably fertilizers and irrigation. Plant breeding has also been effective in improving specific characteristics that have a high level of qualitative genetic control, such as single gene controlled disease resistances. The comparative advantage of modern improved varieties over local

---

* Jaap Hardon and Walter de Boef dedicate their paper to all farmers whose knowledge and capacity to maintain and develop plant genetic resources has been ignored for so many years by crop researchers.

landraces tends to depend on simultaneous manipulation of the environment, through adapting growing conditions to the requirements of the new varieties in order to achieve higher yields. Where opportunities for the use of external inputs do not exist, the comparative advantage in yield of modern varieties over landraces tends to be reduced.

This paper examines constraints in modern plant breeding techniques, and the role farmers play in crop development. It is argued that co-operation between both systems is essential and is in the interests of both farmers and plant breeders. Crop improvement in the formal and informal systems is seen as lying on a continuum, rather than constituting independent, conflicting and competing approaches. Most studies on local knowledge in agricultural technology development have tended to concentrate on how the attitudes and practices of farmers can be incorporated in setting research objectives for institutional research. In contrast, the present paper focuses on the application of modern science to the improvement of local crop development within a research setting in which the basic integrity of the informal system is maintained. Issues concerned with *ex situ* (in genebanks), *in situ* (in the wild or agro-ecological system) and farmer/community conservation of genetic diversity are discussed in the overall context of crop development.

Constraints in modern crop breeding

Agricultural productivity increased substantially when modern varieties were combined with other high input technologies in Europe and North America. Farmers increasingly relied on industrial inputs to raise productivity and control the vagaries of the environment in soil fertility, biotic stresses, and in the most extreme form controlling the total environment in greenhouses. For the first time in European history food production was ensured. Food could be provided at reasonable prices to a growing number of people working outside the agricultural sector, a necessity to sustain industrial development. In the temperate, benign and agriculturally favourable environments of Europe and North America, the technology apparently worked. But second generation problems, including excessive use of chemical inputs and monocultures, have resulted in environmental degradation, a major concern in present European agriculture.

Industrial development and the need to feed a growing world population necessitate increased agricultural productivity world wide. However, there is growing evidence that the present 'short-term' objectives of yield increase in fragile environments may in the long-term irreversibly affect the natural resource base. Product-oriented agricultural technology development, primarily driven by market forces and dependent on external inputs, may also be less effective in solving the more complex problems of resource-poor farmer households in less favourable production environments and under more marginal socio-economic conditions (Biggs and Farrington, 1991). Technology development is faced with the dilemma of balancing short-term macroeconomic needs for more food, with long-term problems of sustainability in relation to conserving the natural resource base and the livelihoods of small farmers.

Modern plant breeding and agricultural research have mainly focused on major staple food and cash crops (cereals, grain legumes, oil crops, a few vegetables, and a number of industrial crops). These crops are adapted to favourable production environments and are cultivated within High External Input Agriculture (HEIA) systems (Reijntjes *et al.*, 1992). Little research is carried out on minor or under-utilized crops (vegetables, legumes, fruits, trees and medicinal plants which are restricted to specific regions). There is also a lack of research on the adaptation of many major crops to more ecologically marginal and diverse environments. In these environments, minor and major crops are often cultivated in Low External Input Agriculture (LEIA) systems (Reijntjes *et al.*, 1992). Intensified production and use of local resources with few or no external inputs in these systems lead to the degradation of natural resources.

Modern plant breeding is limited in its ability to address the diversity of crops and environments by institutional, technical, economic and conceptual factors. Institutional constraints include limited capacity to address the multitude of different crops, environments and cropping systems involved. Technical constraints include a poor understanding of the genetics of tolerances and adaptations to environmentally induced complex biotic and abiotic stress factors, and difficulty in handling these factors under experimental conditions. Yet these factors are a pre-condition for successful performance of crops under LEIA. The economic costs of plant breeding programmes determine that single varieties should have a wide usage, which limits investments in minor crops. Conceptual problems relate to the fact that the whole complex of technology development within the public and private sector is largely product oriented and fails to consider adequately the context in which new technologies have to perform.

Formal and informal systems

As a result of these constraints, two independent systems of crop improvement continue to exist side by side in most developing countries, with their own exclusive linkages for the management of plant genetic resources (Berg *et al.*, 1991; Keystone Center, 1992). This includes:

o A formal institutional system involving international agricultural research centres (IARCs) and other regional and commodity-oriented centres, national research systems and private industry, linked to farmers through extension services and marketing in a linear model of technology development and transfer. The emphasis is on maximizing yields and use of external inputs with the main objective of solving macroeconomic and national needs for more food.

o An informal system at the farm level using and developing crop diversity through local landraces and seed production and distribution at the community level. The major emphasis is on yield stability, risk avoidance and low external input farming.

Strategies for conservation and development of plant genetic resources

Controversy and misunderstanding exists in conceptions of the relationship between formal and informal management of genetic resources. This section discusses some of the issues at stake in conceptions of plant breeding, local crop development and conservation.

Plant breeding and local crop development

Plant breeding, the applied science of genetics, aims to develop new, widely adaptable varieties which satisfy a narrow set of breeding objectives. It is involved in a continuous process of crop improvement by fixing desired characters. The expression of individual characters is maximized in uniform varieties, a requirement further stimulated by the needs of formal seed supply systems and plant breeders' rights.

The objectives of local crop development are basically different from formal plant breeding. Specific adaptation is considered important as a logical consequence of local selection. In addition maintenance of a certain degree of variation within and between local landraces is considered beneficial as a buffer against temporal and spatial variations in biotic and abiotic stress factors in LEIA systems. The need for diversity between landraces in farmers' fields is further stimulated by variations in the use of different landraces by households, for cultural and religious purposes, festivals, etc. In many regions, crops are still an integral part of the local culture (Richards, 1985; Brush *et al.*, 1981; Boster, 1985; Cromwell, 1990).

The term 'local crop development' is proposed instead of 'local crop improvement', since in many instances the objective is not to realize specified improvements but to engage in more gradual processes of adaptation and change. In many situations this is more akin to a process of maintenance than actual improvement. The term 'evolutionary breeding', as used by Berg *et al.* (1991) may also provide a satisfactory term. However plant breeders could claim that breeding merely allows them to speed-up evolutionary processes and is therefore not basically different.

Formal and informal activities are not in conflict with each other but rather represent complementary activities, with their own relevance in time and space. Increasing recognition of the value of local plant material in plant breeding will result in more suitable varieties and facilitate more widespread adoption. However, economic constraints will limit the scope of breeding programmes to major crops and relatively broad environmental domains. Gaps in research will continue to exist for minor crops and adaptations to more extreme, marginal and restricted regions. Under these circumstances local crop development will continue to play a role and should be recognized as part of the overall strategy to develop better adapted material and support for farmers' activities.

Strategies for conservation

Ex situ conservation refers to the collection and long-term conservation of genetic diversity in genebanks. The objective is to sample a reasonable part of the existing genetic diversity (both genes and genotypes) and ensure its availability for future use. New genotypes (varieties, landraces, wild species) can be added continuously as they evolve. Hence the objective is to provide added security and complement maintenance of genetic diversity as it occurs in landraces, modern varieties and wild species. The need of this is amply demonstrated by on-going genetic erosion, both in farmers' fields and in natural habitats.

In situ conservation generally refers to the conservation of wild species in their natural or original habitat. The need for *in situ* conservation of crop genetic resources and the environments where they occur has been emphasized, since *in situ* conservation allows for continued, dynamic adaptation of plants to the environment (Nabham, 1985; Prescott-Allen and Prescott-Allen, 1982; Wilkes, 1983). This conservation strategy may be particularly important in areas under traditional farming, where crops are often enriched by gene exchange with wild or weedy relatives (Harlan, 1965). On the negative side, it is argued that in the absence of controlled monitoring, security is low. Natural habitats are lost, landraces replaced by other varieties, and foreign genetic material is introduced into local landraces as an integral part of farmers' strategy to maintain and improve local germplasm (Wilkes, 1977). A rigorous programme of *in situ* conservation of landraces would require a return to or preservation of 'primitive' or 'original' agricultural systems, which to many scientists, conservationists and development workers would be an unacceptable and unworkable proposition (Ingram and Williams, 1984).

Increasingly, it is argued that conservation and management of plant genetic resources at the local farmer and community level should be linked with development actions (Altieri and Merrick, 1987; Brush, 1991). This strategy has been recognized as the third conservation strategy, complementing both *in situ* and *ex situ* conservation (Keystone Centre, 1991). This conservation strategy can be seen as an integral part of local crop development, maintaining and stimulating the dynamics of the management of plant genetic resources by farmers and local communities.

Sustainability and local cropping systems

The inherent sustainability of local farming systems is often assumed. However, historical analysis does not support these assumptions. There are numerous examples which suggest that whole civilizations collapsed because agricultural productivity, based on traditional knowledge systems and without access to external inputs, was not able to keep pace with a need for surplus food to feed growing strata of people not directly involved in agriculture (Pointing, 1991). In Mesopotamia, the Indus valley, the tropical forests of Mesoamerica, China, the Mediterranean and other areas, pressure for more food has historically resulted in irreversible over-exploitation of natural resources (Lawton and Wilke, 1979; Weiskel, 1989). In modern times many traditional farming systems are constrained by increasing population pressures. Systems of soil regeneration based on shifting cultivation and fallowing are not sustainable under increased population growth. Exceptions which have survived into modern times, such as agricultural systems adopted by the ancient Egyptians in the Nile valley and wetland-rice farming systems, have depended on unique opportunities to exploit natural processes (the flow of water and silt to fertilize the land) to realize nutrient recycling necessary for surplus production.

The dilemma of producing more food while conserving the natural resource base is complex. Neither modern technology nor traditional agriculture have fully come to grips with this problem. It remains the main challenge for agricultural development.

New research perspectives

The quest for both increased production and sustainable cropping systems may partly be solved by developing a complementary relationship between local crop development and modern plant breeding. This requires an interaction between the knowledge of farmers on local crop development practices, and the scientific approaches of researchers, plant breeders and conservationists.

There is a growing literature describing conservation and use of genetic diversity in traditional agricultural systems (Altieri and Merrick, 1987; Brush, 1991). Although such work provides important information, it remains largely empirical and anecdotal. Research is needed to generate analytical data on local crop development. This data will be essential to validate suggested comparative advantages and may provide new options for conservation of plant genetic resources and for plant breeding. This may lead to new approaches which provide better support and improved management of plant genetic resources and crop improvement systems at the local level, while maintaining the inherent characteristics of these systems.

The main areas in which more research is needed include:

Selection criteria. There is only scant data available comparing the performance of modern varieties with that of local landraces on farmers' fields, under farmer management and utilising farmer evaluation. Hence the factors which result in farmers preferring local landraces over modern varieties are not very well understood. The available information suggests that modern varieties often lack additional characters which farmers consider important (e.g. storability, taste). Linked to this is a need to better understand how farmers select material for the following year's crop, their selection criteria and, most importantly, the effectiveness of this selection (e.g. maintenance breeding versus breeding for distinct if gradual improvement in limited sets of characters). The methods which farmers use for field testing, data recording, and the conscious utilization of techniques such as introgression and hybridization need to be researched and documented. In modern agriculture, varieties are seen as an external input, whereas in the concept of low external input and sustainable agriculture (LEISA) the landraces are seen as an integral part of the cropping system. The significance of interactions between cropping and other activities (animal husbandry, off-farm employment) for local diversity needs to be studied.

Local specificity. It is often suggested that specific local adaptation of landraces is important, particularly in more marginal extreme environments in LEIA systems. This may be a logical consequence of local selection. There is evidence of local adaptation where specific climatic factors prevail over long periods (temperature with altitude in the Andes for potatoes, and in Nepal for rice). However, where variations between years in factors such as the onset of rain, and differences in field sites (e.g. variable soil conditions in water holding capacity, soil nutrient status) characterize marginal agricultural environments, local selection may well favour a degree of broad adaptability. An emphasis on

69

yield security rather than maximum yield could additionally favour broad adaptability rather than local specificity, especially when desired secondary characters prove to be less environmentally sensitive. A better understanding of these issues is essential, since these factors determine the range of environments between geographically-isolated localities that may benefit from the exchange of landraces to broaden the genetic base for local crop development.

Abiotic and biotic stress factors. It is generally assumed that genetic diversity within landraces and between crops in cropping systems provides a form of natural protection against both biotic and abiotic stresses. However, more information on such constraints would help in identifying genetic materials, in the field and in genebanks, which can be made available to farmers for direct use or utilized as a source for introgression in local landraces. Local knowledge systems of maintenance of crop diversity, seed production and storage need to be investigated, assessed and compared with modern plant-breeding methods.

The above issues are relevant to both major and minor crops. At the local level, more attention needs to be given to socioeconomic issues such as gender, household economy, self-sufficiency, and market-oriented income-generation.

Building linkages

While modern plant breeding is mainly product oriented, local crop development is involved with a more complex process. It is sometimes argued, notably by some non-governmental organizations, that local crop development has little to gain from institutional plant breeding in the formal system. Past experiences with modern plant breeding, which has aggressively promoted modern varieties without adequate on-farm testing. Land which has failed to elicit farmers' crop requirements and involve them in setting breeding objectives, may give grounds for these views. However, it would seem that at least part of this conflict is based on a lack of appreciation of the inherent characteristics and limitations of both systems of crop improvement exacerbated by wider political and socioeconomic arguments on both sides. On the one hand, plant breeders have assumed that modern varieties bred for improved yield potential would have more general relevance over, in their view, more 'primitive' landraces. Low adoption rates in notably the more marginal production environments were ascribed to inefficient extension and seed production services and generally to a low level of overall development. It has also been claimed that farmers reject modern varieties, because they do not suit their requirements, and give considered preference to local landraces.

It has been argued here that the formal institutional system should recognize and accept local crop development as a genuine system with comparative advantages under certain circumstances which can complement institutional plant breeding. It is suggested that local crop development can benefit from modern genetics and methods of plant breeding. However, the provision of support requires systematic and structural knowledge of local crop development based on genetic theory and the development of appropriate local breeding methodology. Considerable groundwork is provided by recent advances in participatory research methodology in agricultural technology development. This research base can be naturally extended into plant breeding activities since support for local crop development is closely linked with farming systems and their socioeconomic and household context. The introduction of participatory methods in genetic resource development will break new ground, benefiting both formal and informal plant breeding. It will also clarify the role of small farmers in plant genetic resource conservation, through maintaining agro-ecosystems in which these resources occur.

In the further development of strategies to support local crop development, it is suggested that co-operation be established with on-going grassroots rural development programmes in a variety of natural and socio-economic environments, in different cropping systems utilizing different crop mixtures, and in farming systems reflecting differing emphasis on market-oriented production and household food self-sufficiency. A broad and integrated approach is necessary to provide information which will lead to more relevant approaches to crop improvement in a greater range of regions and environments.

The science of plant breeding — support or alternative to traditional practices?

Trygve Berg[*]

The comparative advantages and constraints within systems of traditional and scientific plant breeding are examined. While traditional plant breeding is constrained by its technological base it has developed sound strategies in breeding for complex and diverse environments. The selection normally produces heterogeneous varieties with specific adaptation. Scientific breeding, however, has been influenced by market factors which determine that seeds should be broadly adaptive and uniform. This results in breeding processes which are characterized by cloning and intensive selection within populations, but which may not be sustainable in the long term, particularly in marginal environments. Processes involved in seed selection in southern Sudan and Tigray are described and the potential of utilizing community organization for improved seed selection examined. It is argued that enhanced germ-plasm can be supplied to local communities to strengthen their breeding programmes.

'Plant breeding is a science, an art, and a branch of agricultural practice'. With these words Vavilov (1951), the pioneer of modern genetic resources exploration, opened an essay on how an old practice could become a science through the application of the newly discovered principles of genetics. Since Vavilov's time the science of plant breeding has become established as a function of the formal seed supply system. The traditional farmer practices of plant breeding have tended only to survive where the formal seed supply system has failed to dominate. Thus modern and traditional plant breeding have become associated with different seed supply systems and are evolving independently. The majority of the breeders of the formal sector are unaware of the interest in agricultural practices which were Vavilov's point of departure.

Local plant breeding activities are considered inefficient by most authorities and development agencies, and the communities where plant breeding is still practised are seen as targets for the introduction of formal seed systems. This paper examines the characteristics, constraints and comparative advantages of both types of plant breeding, and the types of linkages which would enhance local plant breeding activities.

---

[*] Trygve Berg dedicated his paper to Gebremedhin Kidane of Mai Brazio in Northern Ethiopia.

Traditional and modern plant breeding

The differences between modern and traditional plant breeding cannot be defined in terms of technique and efficiency. The strategies of plant breeding activities are influenced by different aims and objectives. Thus the technology and the strategy of plant breeding need to be disentangled (Table 2.1).

Traditional communities may benefit from more efficient breeding technology. But they may not benefit from mimicking the strategy of scientific breeding, which has been conditioned by the requirements of the formal seed supply system for broad adaptive qualities and uniformity in modern varieties for mass dissemination. From the individual farmer's point of view, specific adaptation and variability are normally advantageous. Another point is that traditional plant breeding also may be the only sustainable genetic resource management system. There are strong reasons for exploring the feasibility of using science to strengthen indigenous breeding practices. Specifically, this means combining the use of modern techniques with the strategies of traditional systems.

Specific adaptation and variability

The activities of seed companies are defined by the need for varieties which can be used by as many farmers as possible. In contrast, the traditional seed selector is concerned with the performance of varieties in specific localities. The process of selection and utilization of genetic material within specific localities is less complex than the selection of materials to perform over a range of widely scattered locations. The latter strategy involves compromising the best local/specific adaptation.

A geographically-wide adaptive sphere is feasible when varieties are bred for standardized cultivation methods. In areas with highly complex and varied farming systems it is much harder to achieve a site adaptation of breeding materials. In these areas local selection would be a more feasible solution.

It is commonly assumed that intensive selection within a population leads to reduction in genetic variability and ultimately to erosion of the basis for further selection responses. According to this assumption, uniformity should be an inevitable result of scientific breeding. However, within traditional communities high genetic variability exists despite thousands of years of seed selection. This assumption is also contradicted by modern selection experiments where experimental populations under intensive long-term selection have failed to show signs of diminished genetic variability. In a renowned experiment with maize, it was found that after 53 generations of selection for high and low oil and protein content, the selected strains were still diverse enough for both continued and reverse selection. Selection responses were still high after 76 generations of selection (Dudley, 1977).

Single genes for qualitative characters can be fixed or lost if they are put under a strong selection pressure. But most important characters are polygenic: they are governed by a high number of genes, each with a small additive effect. The real variation in characters will always be much less than the theoretical range. The genetic potential implies possible changes far beyond the range of variation in the original population. Therefore long-term selection can be exercised with sustained selection responses. Within a population under selection,

Table 2.1: Components of plant breeding technologies and strategies

Plant breeding activity		Scientific breeding	Traditional breeding
Breeding Technology	Genetic resource base	World genetic resources	Local genetic resources
	Crossings	Controlled	Random
	Selection method	Efficient	Moderately efficient
Breeding Strategy	Adaptation	Broad adaptation	Specific adaptation
	Variation	Uniform varieties	Heterogeneous varieties

gene frequencies will change but the different versions (alleles) of each gene may survive. In the above mentioned experiment, selection was estimated to have raised the frequency of high protein alleles from 0.37 in the original population to 0.62 after 48 generations. Selection response is thus achieved through higher frequencies of favourable alleles within the plant population and not through the fixation and loss of alleles. Potential genetic variability will therefore be sustained given that the population under selection is large enough to escape inbreeding and genetic drift. This does not mean that there is no limit to selection response but that with quantitative traits such as yields, which are governed by a high number of genes, the theoretical limit is far beyond that reached in any practical breeding programme.

Thus, theoretical population genetics can explain what has happened under traditional long-term seed selection. It has changed the characters under selection far beyond the range of variation within the original wild progenitors. But the potential variation has also been sustained, allowing for both reverse selection towards the original types or continued selection towards more extreme types. In addition to the long-term improvement of quantitative traits, traditional selection has occasionally added new qualitative traits which originate through mutations. In these cases the beneficial alleles may be fixed.

There is no genetic reason why this 'evolutionary breeding' could not continue with the application of more sophisticated methods of selection. Formal plant breeders emphasize uniformity because the commercial market demands stable and distinct varieties. This is achieved through inbreeding or cloning, and not through intensive selection within populations. In contrast, in livestock breeding where the market in many cases does not demand stable and distinct varieties, evolutionary breeding is applied using the most sophisticated selection methods. In animal breeding schemes, the selection response may be high without reducing the potential genetic variation of the breeding population.

The scope for improvement in local breeding technology

Local crop varieties, which have been maintained and used by farmers, have been systematically collected by scientists for utilization in breeding programmes. Local breeders, however, remain confined to their traditional seeds. While the diffusion of varieties through informal networks of neighbourhood exchange may add new germplasm to local genetic resource bases, this does not allow local breeders to explore the potential of knowledge emanating from scientific institutions and to carry out their work on equal terms with scientific breeders. The dissemination of genetic materials in the form of enhanced germplasm for local selection could be a way of compensating traditional seed selectors for their contributions to the genebanks of the world.

Traditional seed selection methods

The scope for improvement of traditional selection skills depends on certain aspects of seed selection. This includes the degree to which seed selection is considered important, the people involved in breeding, time allocated for seed selection, and the methods through which knowledge of seed selection are transmitted to the next generation. The problems involved can be illustrated by experiences from Southern Sudan during the early 1980s, before the current civil war, and from a recent post-war visit to Tigray in Northern Ethiopia.

In Southern Sudan, I discovered local plant-breeding activities by accident. While participating in a germplasm collection mission, we came to a village where, after some discussion with the people, we thought we had been granted permission to take some heads of sorghum. But on picking the sorghum a woman came shouting furiously after us. Eventually we found out that, as the mother of the family, she was responsible for selecting the supply of seeds. Every year she would select the best sorghum heads from the field before the harvest could start. Removing seeds before she had made her selection was forbidden. We had violated this prohibition.

Through this incident it became apparent that there was a strong culture of plant breeding, that the breeders were women, and that selection was done in the field immediately before harvest. However further investigation revealed that the process was more complicated. Selection is based on observation during the growing season and all the farmers may be involved in assessing the performance of the crops.

During the ripening period, boys are posted in the field as bird scarers, keeping a watchful eye on the sorghum heads and chasing away intruding birds. They are overseen by their fathers who check on them from time to time and survey the whole fields, examining the sorghum heads for signs of bird damage. During the same period, women and girls regularly come to the fields to gather intercropped vegetables and edible weeds. They also carefully observe the sorghum plants, looking for candidates for selection. By the time of collection of planting material for the following season, the women already know from a long period of observation and from family discussions the best sorghum plants.

During a visit to Tigray in the period of the main cereal harvest in November 1991 a similar picture emerged. I travelled for three weeks to villages all over the region to review a community seed bank project (Berg, 1992). The seed banks supply seeds on credit, and utilize experienced local seed selectors identified by the community to make sure that their seed stocks are of the best quality.

In discussions with these seed selectors one of them commented that the farmer who selects seeds at harvest time is lazy. The normal procedure is to observe the performance of the crop from germination and throughout the entire growing season, and discussions are regularly held within the family and neighbourhood on the performance of seeds. The performance of seeds is a common subject of conversation as farmers walk through cultivated fields. They also watch the fields of their neighbours and may request seeds they are interested in planting.

These observations reveal the existence of a culture where seed selection is supported by great knowledge, interest, discussion and the devotion of much time to these activities. Children are always involved, and the transmission of skills to the next generation is emphasized. It is likely that such communities would respond with great interest to offers of training in improved selection skills, if they understand that this will help them achieve their own objectives.

Organizing local selection: the Tigrean model

Traditional seed selection activities cannot easily receive external technical support unless they are organized. Local varieties evolve as community germplasm through the collective efforts of a network of people. The organization of local seed-selection activities must take the form of a community undertaking.

During the recent war in Tigray, plant breeding societies were organized in certain areas which were isolated from governmental services (Berg, 1992). This started as a community response to famine in the mid 1980s. Traditional seed selection was declining, but those who practised the traditional skills were getting better yields of crops. Farmers who lost their seeds during the drought were also exploited by seed lenders who charged exorbitant rates of interest. As a result Community Seed Banks were organized to extend the merits of traditional seed selection and to provide non-exploitative credit for seeds.

The seed banks are owned and managed by the *baito*, an elected body at the sub-district level. The *baito* identifies traditional seed selection experts whose skills are utilized in purchasing prime seed. The selected seeds are kept by trusted female seed keepers. Loans are granted by the *baito* and the borrowers receive the seeds at planting time for the price which the *baito* paid when it was purchased. The borrowers repay the loans after harvest with between 6 and 9 percent interest. In cases of crop failures the *baito* may accept a one year delay of repayment without additional interest. The initial purchase of seeds was financed through funds granted from external sources by NGOs and the money is utilized as a revolving fund. There seems to be strong social pressure for repayment and most of the seed banks have managed to maintain their initial capital fairly well. Seed selectors and seed keepers provide their services free of charge. The administrative tasks are shouldered by the *baito*.

The seed banks are popular and successful because they are managed by the communities, and because the advantages of the services they provide are clearly visible. This type of community organization could form the necessary base for provision of external technical support for the community seed system.

Conclusion

Scientists often approach traditional societies with the aims of extracting valuable materials and without thought to the inputs they can make to these societies. Plant breeders take genetic resources and social scientists collect indigenous knowledge and other information. The scientists need to be challenged to reverse the flow of valuable resources and information. Genetic resources, in the form of enhanced germplasm, should be given to local plant breeders in order to broaden the local selection base. Scientific knowledge needs to be made available and accessible to local seed selectors and producers in order to increase the efficiency and quality of traditional productive activity.

Linking genetic resource conservation to farmers in Ethiopia

Melaku Worede and Hailu Mekbib*

Farmers in Ethiopia have a wide knowledge of crop varieties and have developed strategies to preserve diversity in crops and genetic material. This enables crops to be adapted to specific environments and to conditions of stress. Given the weakness of modern varieties in performing under conditions of stress and in adapting to specific environments, farming communities can play an important role in genetic resource conservation and development. Plant Genetic Resource Centre/Ethiopia programmes, in which farmers play an important role, are described. Improved landrace conservation will enable farmers to exercise greater choice in adopting planting material and in rejecting poorly-adapted exotic varieties.

The knowledge of farmers is an important resource for the development of sustainable agriculture and the conservation of genetic material. Farmers have developed landraces on the basis of utilizing locally accessible resources for the management of farming systems. The indigenous landraces of various crop species and their wild and weedy relatives form the basis of Ethiopia's plant genetic resources, and are highly prized for their potential value as sources of material for crop improvement programmes. Genetic resource development forms an important activity in Ethiopia and major national programmes have been undertaken over the last decade. Crop scientists are presented with the challenge of developing new approaches to the country's rich and diverse biological resources. There is a unique opportunity to salvage and effectively utilize landraces and indigenous knowledge which farming communities in Ethiopia have developed and maintained through the centuries.

This contribution reviews and describes the importance of local knowledge in traditional agricultural systems, and discusses the role peasant farmers play in the conservation of landraces and in the programmes of the Plant Genetic Resource Centre/Ethiopia (PGRC/E). It is important that the dynamics of traditional cropping systems should be understood before they are replaced with modern agriculture, and that peasant farmers should be supported with resources to become partners in the development of genetic resources.

---

* Melaku Worede and Hailu Mekbib dedicate their paper to the farmer Ato Husen Yimer Hassan, who plays a great role in the on-farm conservation in Wollo Province, Ethiopia.

Crop diversity in farming systems

Farmers frequently grow mixtures of different crops which are adapted to different localities. This is done to reduce the risk of economic loss caused by undesirable environmental conditions or pest attack. The small farmers are more concerned with gaining maximum security from their cropping systems than with maximizing yield, to assure that the basic food requirements of their families are met.

In northern Ethiopia, particularly in the drought prone areas, wheat and barley are grown in particular mixtures. In favourable years farmers will get yields of both crops, and in poor years they will mainly reap barley. The mixture of landrace populations consists of genetic lines, which complement each other. They are all adapted to the region in which they have evolved, but differ in the mechanisms through which they express traits such as drought or pest resistance. The mixtures of both crops are kept together for the coming planting season, but during consumption the two crops are used separately for different food preparations.

In the Gonder area of northwestern Ethiopia, farmers plant more than six crops together in their backyards, including maize, faba bean, sweet sorghum (used for chewing the stalk like sugar cane and for chicken feed), cabbage, tomato, potato, pumpkin and bottle gourd. Most of these backyard activities are the responsibility of women. In the southern and central part of the country the farmers focus more on perennial crops. A highly diversified range of crops and trees used for fencing materials are planted (Figure 2.2). These crops mature at different periods making maximum use of scarce land and labour resources, minimizing weeding problems and maintaining soil fertility.

Exchange of seeds and planting material

Farmers have developed networks and systems of ensuring a sustained supply of seeds. Seeds are exchanged in local markets, where an assortment of varieties adapted to different environmental adaptations are available. Inter-regional exchange is also important and farmers know where to locate new supplies of seeds when traditional landraces become degraded.

In northern Ethiopia, in areas where lands are flat, wind and water can easily carry pollen from one field to another. Under these conditions farmers find it difficult to maintain traditional landraces and visit other areas every few years to acquire seeds of recognized landraces. Farmers in Kanesham in Eritrea are able to maintain distinct landraces of barley on small and isolated plots. Areas in Gonder and Tigray also specialize in the maintenance of 'elite landraces' of other crops. It is important that other such areas are identified and studied and that the knowledge of farmers sustaining the processes which produce elite landraces is recognized. These practices enable farmers to have a wide choice of planting material suited to particular agro-climatic conditions.

In some of the more developed areas of Ethiopia, such as the Central Highlands, farmers' traditional seed conservation activities are becoming eroded as new improved seeds are spread by the extension systems. But in most of the drought prone areas, particularly in northern Shewa and Welo, farmers depend on traditional methods of seed dissemination and production to ensure a supply

Table 2.2: Crops planted in Southern and Central Ethiopia

Name	Scientific name	Name	Scientific name
Enset	Ensete ventricosum	Ruta chalepenisis	Ruta gravelcens
Rhamnos	Rhamnus prinoides	Dama kese	Ocimum lamifolium
Sacred basil	Ocimum sanctum	Ariti	Artemisia afra
Chili	Capsicum frutescens	Tej sar	Cymbopogon citratus
Potato	Solanum tubersum	Pumpkin	Curcurbita species
Lemon	Citrus aurantium	Kacha	Hibiscus canabinus
Maize	Zea Mais	Sugar cane	Saccharum officinarum
Sweet potato	Impomoea batatas	Astenegar	Datura stramonium
Garlic	Allium sativum	Red shallots	Allium ascalonicum

of adaptable planting material.

Mechanisms have also been developed for the storage of seeds. Individual farmers often store seeds in clay pots and rock-hewn mortars, which are sealed, buried or stored in other secure places. Community grain pits are also made in which different crop seeds are stored. The grain pits are carefully prepared and fumigated with smoked, dried cow-dung or wood from selected tree species.

Farmer selection and maintenance of seeds

Farmers expert in traditional seed selection have great knowledge of varieties. The traditional criteria for selection of landraces include induced adaptability, high yield, reliable and stable yield, nutritional quality, colour, grain size, and texture. Crops are also adapted for specialized uses. In the case of wheat varieties, *Triticum aestivum* is used for making common bread, while *T. turgidum* conv. *durum* and *aethiopicum* are used for macaroni, spaghetti, pastries, local breads, whole and crushed grain foods, drinks, and porridge. *T. turgidum* conv. *dicoccon* is used to prepare a soup which is consumed by women during pregnancy and weaning. Various alcoholic drinks are brewed and distilled from a variety of wheats.

All these recipes are prepared by women. Thus women's role in seed selection and vegetative propagation is crucial to both agricultural production and the conservation and enhancement of genetic resources. Women have traditionally played a silent but central role in the sustainable use of biological resources and life support systems (Shiva and Dankelman, 1992).

The maintenance of species and genetic diversity in fields is an effective strategy to create a stable system of conservation by farmers practising low input agriculture. Cultivated crops often intercross with their wild or weedy relatives growing in the field or in nearby fields, resulting in new characteristics. This has been observed for ensete (*Ensete ventricosum*) in central and southern Ethiopia. Farmers have taken advantage of this system of interbreeding to adapt materials to changing agro-ecological realities. Strategies of intercropping and cropping with varietal mixtures may result in accidental crosses between the varieties promoting introgression. Bayush (1991) reports that farming communities in southern Ethiopia actively manage germ-plasm through artificial selection. It is believed that the different characteristics of plants within the brassicas, such as *Brassica carinate* (Ethiopian mustard) and *Brassica nigra* (black mustard), arose on farms in which mixtures of these two species were planted. These practices may also account for the relatively high interpopulation diversity which has been observed in PGRC/E collections (Worede 1986a). Coffee growers often preserve a diversity of local varieties in small areas alongside the more uniform coffee blight disease resistant line which is distributed by the coffee improvement projects in Ethiopia.

The PGRC/E has benefited from the knowledge and skills of farming families collaborating in genetic resource activities, especially in collecting and rescuing germ-plasm and in the identification of useful planting material. This has already contributed to available information on Ethiopia's crop germ-plasm resources which farmers have developed and maintained for many generations.

The role of farmers in genetic resource conservation

Given the knowledge and skills within the traditional system, conservation of landraces on peasant farms provides a valuable option for conserving crop diversity. It increases the range of strategies for genetic resource conservation efforts (Worede, 1986b), and provides a mechanism through which the evolutionary systems responsible for the generation of variability are sustained. In relation to pests and diseases this will allow continued host-parasite co-evolution. Within stressful environments, access to a wide range of local landraces may provide the best available planting strategies. The ability of landraces to survive under such stress is conditioned by an inherent broadly-adaptive genetic base. This is often not the case with the more uniform improved varieties which, despite their high yield potential, are less stable and hardy under adverse growing conditions.

Under these conditions, the establishment of field genebanks of species adapted to extreme-conditions may provide a seed reserve for post-drought planting in places where traditional crops have failed. Germ-plasm materials maintained in such fields could be distributed to rural farming communities, scientific institutions and other organizations for further investigation on potential use as food resources and for utilization in plant breeding programmes.

Landrace evaluation and enhancement programmes will stimulate the utilization of germ-plasm resources which are already adapted in these regions to the conditions of stressful environments. Local landraces will provide suitable materials for institutional crop improvement programmes, but need to be maintained within the dynamic conditions which have characterized their evolution and selection, within farm or community-based conservation programmes.

PGRC/E's farm-based conservation activities

Work has recently begun in Ethiopia to develop farm-based conservation activities, which build upon the skills and traditions of farmers. The PGRC/E is presently involved in developing three programmes with farmers.

On-farm landrace conservation and enhancement

Since 1988, farmers, scientists and extension workers have been involved in a programme of genetic resource conservation in northeastern Shewa and southeastern Walo, with support from the Unitarian Service Committee of Canada. The aim of the project is to help farmers maintain crop diversity by protecting cultivars from disappearing, and also by improving their genetic performance. Materials previously collected from surrounding areas and regions are given to farmers to plant and to carry out simple forms of mass selection to improve their characteristics. Farmers are assisted by breeders, and other scientists have access to the farmers' fields for the purposes of carrying out research. Most of the farmers are women and were selected through farmer co-operatives.

Farmer seed selection is usually carried out at the time when crops have formed heads, when the various plant types become discernible. Plant types and subraces of cultivars are selected for such characteristics as disease or pest resistance, size of kernel or head, maturity period, and other characteristics of local importance. The long-established skills of the farmers are complemented by PGRC/E scientists who establish standard descriptor lists (such as ear length, ear width, number of tillers, disease and pest resistance, lodging resistance, etc.), which they train farmers to follow, enhancing their selection skills. Occasionally farmers also rogue out suspect varieties at an early stage for such diseases as smut in sorghum. A small number of plants are identified for each cultivar, harvested and the seeds of the selected stock are bulked to form a new slightly improved population, which the farmers multiply to continue the process in the following season.

Special experimental plots are planted in which farmers evaluate their seeds by comparing performance and yields with samples of the original seed stock. These plots are also used for on-site maintenance of landraces, sampling and other relevant scientific experiments. The traditional cultural and cropping practices under which the plants have acquired their distinctive properties are maintained to optimize conservation. The multiplication and selection of elite materials for further selection is carried out on separate plots. After three to five years of selection, an appreciable improvement in crop yield is to be expected.

There is also a possibility of transferring genes which control particular positive characters (e.g. disease and pest resistance, high lysine in sorghum, drought tolerance, etc.) from external sources, or from already-existing selections to enhance the elite populations of the future. Certain types of cultivars which, despite positive attributes, may have been abandoned by farmers for various reasons, such as risk of crop failure or marketing problems, could also be rescued. Currently, a lot of germplasm which has already disappeared in the surrounding area, due to adverse growing conditions and insecurity and war, has been maintained on these landrace conservation farms.

Maintaining elite indigenous landrace selection on peasant farms

The PGRC/E in conjunction with Debre Zeit Research Centre of the Alemaya University of Agriculture is developing a programme to maintain elite indigenous material of tetraploid wheat. The project involves farmers in multiplying and using elite seed stock provided by breeders and which is best suited to their conditions.

The programme utilizes wheat germplasm collected by PGRC/E over the last seven years. The approach is based on a modification of conventional mass selection. Different genetic lines are selected for their adaptation to specific environmental conditions, such as stress. After yield testing, two or more superior lines are bulked for further multiplication and distribution to farmers (Tessema, 1986). The farmers multiply and use the stock best suited to their conditions. PGRC/E maintains representative samples for long-term storage at their genebank. This enables farmers to experiment with the elite landrace lines without the threat of losing the old indigenous populations which are accessible from within the genebank. They are being encouraged to experiment in order to evaluate critically the present available planting materials which are represented by new, poorly adapted varieties distributed to farmers in the region, and to continue using the elite landraces.

Preliminary yield trials indicate that the selected elite landraces have in many ways surpassed officially-released wheat varieties. There is a potential for breeders to develop programmes which select or segregate superior landrace varieties in order to provide suitable breeding material for specific environments (Tessema, 1986). There is also a possibility of utilizing the various selections for continual hybridization, possibly by employing chemical male gametocides, allowing new, unusual genetic combinations to surface, which would be unlikely to occur under controlled manual-hybridization programmes (Worede, 1974).

At the national level, varieties adapted from local landraces can be released to ensure that farmers have a long-term choice of seeds and can fall back on improved versions of adapted local varieties when high risk seeds fail. This is particularly relevant for areas with marginal growing conditions or extremes of environment, where improved varieties fail to meet the requirements of farmers.

Field genebanks for drought-prone areas

Climatic change within Ethiopia is likely to have a serious impact on crop production. A few areas have already been abandoned due to persistent drought and constant crop failure. In other areas, there is a significant shift to more drought resistant crops. In this situation, research into species adapted to extreme environments is important, as is the creation of field genebanks which will provide a source of drought-resistant seeds. Germplasm materials maintained in these genebanks could be distributed to farmers and scientists for further investigation of their potential use as food resources and in breeding programmes.

There are several wild plants which have the potential of surviving droughts where conventional crops perish. These are commonly known as famine crops, since they provide humans with food in times of drought. Some research has been carried out by PGRC/E with *ye-eb* (*Cordaeuxia edulis*), a perennial bush which grows wild in the Ogaden region with seeds which are used by nomads as a highly nutritious source of food. Its leaves are an important fodder crop and cosmetic dye. The plant thrives on marginal soils in semi-arid conditions of less than 200mm annual rainfall. A field genebank is being developed in collaboration with AUA at Dire Dawa in eastern Ethiopia, which will test famine crops and involve farming communities in maintaining and evaluating seed.

Linking landrace conservation and enhancement to utilization

Landraces are stable dependable sources of planting material which are adapted to local growing conditions. In environments where modern varieties fail to meet the requirements of farmers, the conservation and enhancement of landraces on farms is an important objective. But it makes little sense to conserve landraces unless systems are developed for the multiplication and distribution of seeds. Community seed production, marketing and distribution forms a rational solution to this problem since these can be built onto the pre-existing traditional networks of seed selection. This approach enables farmers to exercise control over the availability of seeds and gives them ready access to locally-adapted planting material. They will be able to critically evaluate the relative performance of a wide range of varieties. These factors will act as checks on the undue expansion of costly and poorly-adapted modern varieties.

Strengthening the informal seed system in communal areas of Zimbabwe

T. Andrew Mushita*

Despite an unfavourable policy environment, farmers in the marginal communal areas of Zimbabwe still continue to select and cultivate small grains, which they have adapted to particular environmental conditions. The factors influencing farmers' selection strategies and the role of small grains in farming systems are described. Modern varieties do not perform well in marginal environments and their cultivation with proscribed inputs often exacerbates degradation. While the formal sector aggressively promotes modern varieties, ENDA Zimbabwe has developed a programme to strengthen farmers' informal seed selection systems, for the conservation of small grains in marginal environments.

In the communal areas of Zimbabwe indigenous farming systems have been marginalized by colonial land laws, land-use planning legislation and agricultural policies. Land policies alienated land from farmers and undermined local farming systems and their capacity to manage the fragile environment. Agricultural policies aggressively promoted new technologies while denigrating indigenous farming systems as 'backward' (Elwell, 1991). The introduction of new monocropping technologies associated with hybrids, fertilizers and pesticides has been largely propelled by profit and the need to develop agro-industries and export-crop orientation. Crop improvement programmes during the pre-independence period in Zimbabwe emphasized export crops, such as tobacco, tea, coffee, maize and oil seeds, and neglected indigenous food crops (Collinson, 1983). The spread of high-yielding varieties in low potential areas has resulted in declining use of open-pollinated varieties which were adapted to the environment, and increasing vulnerability to food insecurity and malnutrition. The improvement in yields which farmers experienced from adoption of hybrid maize was the compensation for the possible decline in nutritional well-being of the people from the decline of pulse and other nutritious crop production.

The adoption of these technologies has resulted in problems of environmental degradation, reduced soil fertility, declining yields, and increasing vulnerability of crops to pests, disease, weed competition and drought. It is now being realized that the adoption and application of imported technologies is 'totally inappropriate to our soils, climate, environment and population needs' (Elwell, 1991).

---

* Andrew Mushita dedicates his paper to Mr. Jonga who has managed to maintain a great deal of genetic diversity of small grains in Mutoka area.

Despite marginalization, the remaining farmers operating within indigenous farming systems have managed to solve problems and to successfully manage the environment in ways which baffle researchers. In the light of the skills within indigenous farming systems, ENDA Zimbabwe has established a project to work with farmers in strengthening their small grain production and distribution systems.

Background to the project

Within the wards of Mutoko, Chipinge, Zvishavane and Plumtree, ENDA-Zimbabwe established a project to investigate small-scale farmer management of genetic resources in which 5000 households participated in the implementation of the different research components through the formation of seed committees. The research activities included:

o germplasm collection;
o identification of crop-variety characteristics preferred by farmers;
o crop improvement;
o seed multiplication, distribution and supply, storage and utilization practices;
o agronomic trials designed to compare the overall performance of traditional and improved crop varieties (sorghum, pearl millet and maize) and responses to fertilizers.

Meetings and workshops were organized with farmer groups to discuss and share experiences related to their planting methods, pest and disease control, seed selection techniques, seed treatment and storage, and local seed exchange practices. The workshops also facilitated exchange of information between farmers. The collection of on-farm crop data opened up potentials for farmers to locate plant genetic material extinct in their own locality, and to gain access to new seeds outside the circuit of capital intensive hybrid varieties.

Traditional farming systems

Research revealed that farmers were still cultivating open-pollinated varieties, despite the extensive promotion and marketing of hybrids. This was related to the fact that the distribution of the staple food crop cultivars was closely associated with different soil and vegetation types, resulting in a careful selection of varieties to match environmental conditions.

Particular agricultural practices were also associated with specific agro-ecological conditions, including the use of crop rotations, intercropping, winter ploughing, soil and water conservation and organic farming (Mushita, 1991). Specific benefits arising out of these practices, which were identified by farmers, included higher yields, more reliable yields, minimizing of risk, pest control and improved soil fertility. Mixed cropping or intercropping, involving a combination of crops with different food values, maturity periods, and capacity to withstand calamities, represented one of the most important practices within traditional farming systems. Intercropping and crop diversification can minimize crop fluctuations and total crop failure, thereby helping to stabilize income and food supply. These practices are complemented by the cultivation of crops with a wide genetic base.

Small grain varieties

In the four areas researched by ENDA-Zimbabwe, farmers cultivated an average of eight sorghum (*Sorghum bicolour*) and five pearl millet (*Pennisetum typhoides*) varieties in each area. The characteristics and attributes of the above crop varieties included earliness, overall agronomic stability, yield potential, drought tolerance, grain colour, palatability and storage quality. Although a narrowing of the genetic base is occurring, which farmers have also observed, the maintenance of a wider genetic base as a risk-averting measure is still considered important.

Farmers cultivate a combination of short, medium and long-season varieties of sorghum and pearl millet in the research areas. The proportional distribution of the crops cultivated according to their growth cycle is as follows:

Sorghum		*Pearl Millet*	
Short season	11.0%	Short season	12.2%
Medium season	68.0%	Medium season	75.5%
Long season	21.0%	Long season	12.2%

Short-season varieties are preferred for their earliness in times of drought. However, they are not ideal when rainfall distribution is spread across the season. Medium season varieties are preferred by the large proportion of farmers because they have a number of desirable characteristics and traits such as yield potential, grain size, drought tolerance and good storage quality. They have the ability and potential to cope with moderate climatic conditions. Long season varieties are usually tall and heavy yielding but require favourable rainfall. They provide food, construction materials and animal feed.

Seed selection

Farmers carry out both pre-harvest and post-harvest seed selection. Pre-harvest selection is practised by most farmers. The criteria for selection include overall agronomic and physical characteristics, disease and pest resistance, head shape (true to type), and grain size. Desirable plants are marked out and harvested separately.

Post-harvest selection is carried out at threshing places. The retained seed is selected from harvested heads. The selected heads are preserved by various methods and techniques, including the mixing of seeds with finger-millet ash, hanging heads in the kitchen where they are treated with smoke, sealing seeds in clay pots, mixing heads with millet to avoid infestation, and mixing seeds with cow dungs.

The cultivation of seeds in mixtures also serves to promote crop genetic diversity. Seed exchange among farmers ensures conservation and utilization of a wide genetic base. Farmers often exchange seeds with farmers in other localities through which they gain access to new crop varieties which have been adapted to similar environments. The new crop cultivars are tested on-farm, evaluated and classified.

Farmers also evaluate the overall performance of any improved material in comparison with their own local crop cultivars. While plant breeders and agronomists focus on yield characteristics, farmers are interested in a number of factors, which are difficult to combine in conventional plant breeding which focuses on producing general varieties which can be grown in a wide range of agro-ecological zones. Farmers prefer cultivating crops with a broader genetic base to allow and facilitate the staggering of labour at various peak periods, to cope with different environmental factors, and for selection of different end-utilization qualities, including palatability, nutrition, food, beer and storage.

Strengthening the informal sector

Crop improvement programmes of food staples have had little success, particularly in the case of small grains in Zimbabwe. Despite frequently promising on-station results, yield gaps of up to 60 per cent are consistently observed when most new varieties are cultivated by farmers (ICRISAT, 1980/83). Unacceptable taste as well as processing and storage problems are also commonly encountered. Matlon and Spencer (1984) state that experience suggests that this poor record is caused by a complex of factors, including excessive emphasis given to the development of high-yielding but input-dependent crop varieties without sufficient consideration being paid to the African region's soils, level of infrastructure development and limited farm-level capital.

The success of new materials is limited, as they are generally developed and selected under research conditions which are not typical of prevailing farmers conditions. The situation is made worse by paradigms which assume that farmers are only consumers of research products and not researchers and creators themselves.

There are some plant varieties which now only exist as a few, limited samples which have been protected on small isolated plots for generations by a few families. Developing a support structure for the local breeding of these plants in the farming systems where they occur may be the only way to slow down the process of genetic erosion and ensure that the varieties survive. Support for local breeding and conservation involves responding to the specific needs of small farmers. New research agendas require that plant genetic resources are conserved at the local level with the participation of farmers. It is important for the formal system to understand the dynamic and active role of farmers in crop development and their long experience in this field.

Sorghum genetic resources of small-scale farmers in Zimbabwe

Saskia van Oosterhout*

The criteria which farmers use in classifying and selecting sorghum varieties are examined. These include gastronomic, early maturity and agronomic criteria. Morphological criteria which have received the bulk of plant breeders' attention are of little importance to farmers. Farmers' selection activities are dynamic and enable new qualities to evolve in farm crops in response to environmental changes. This reveals weaknesses in present concepts of ex situ *conservation, which do not take processes of evolutionary change and adaptation in existing crops into consideration. A participatory framework for research is developed which enables farmers' selection criteria to be incorporated into research, and research to be carried out within existing farming systems.*

A key element in the survival of small-scale farmers has been their access to a rich and varied genepool, which has been selected and built up over the centuries. However, in Zimbabwe, contemporary scientific breeding efforts have focused on green revolution technologies and small-scale farmers have been encouraged to concentrate on a few high-yielding varieties. This has led to a progressive narrowing of the genetic resource base of indigenous food crops. A shortage of information about the attributes of indigenous genetic resources has meant that the sorghum improvement programmes have made almost exclusive use of foreign germplasm obtained from the CGIAR system.

Information on rural peoples' knowledge of indigenous crop varieties and the farming practices employed in their cultivation serves to indicate areas where rigorous scientific testing can substantiate the advantages of a particular management system. This paper presents data on the varieties of sorghum grown in the two study areas. The aim is to highlight the choices made by small-scale farmers and to build upon the existing wealth of genetic diversity by providing recommendations which are more appropriate to the conditions under which small-scale farmers grow and select their sorghum varieties.

Methodology

Research was located in Musikavanhu, Gokwe and Siabuwa Communal Areas, resource-poor regions in Zimbabwe with extremely harsh environmental

---

* Saskia van Oosterhout's paper is specifically dedicated to the men and women small-scale famers of Zimbabwe for their vitality and persistance against all the odds of the harsh environment they live in.

conditions. Temperatures are high and rainfall is low, variable and unreliable. The soils are leached granite-derived sands or heavy clays derived from basalts.

Data was collected through a questionnaire survey of randomly chosen farmers. Farmers were also asked to describe and evaluate their sorghum varieties in semi-structured interviews. Care was taken to minimize interviewer influence on the descriptions used by the respondents. The identification of varieties was verified in the field. Varieties were consistently referred to by a specific name within a given area. Answers were classified into three classes, related to agronomic, gastronomic and morphological descriptions.

Farmers' sorghum varieties

A large number of sorghum varieties were grown by farmers, ranging from eight in Siabuwa to ten in Musikavanhu and thirteen in Gokwe. This is typical of the range reported for other areas in Zimbabwe (A. Mushita and J. Lynam, pers. comm. 1992). A maximum of four varieties per farmer were grown in Gokwe, seven in Siabuwa and up to eight in Musikavanhu. However, most individual respondents grew about three to six varieties in Musikavanhu and two or three varieties in Gokwe and Siabuwa. *Chidhomeni, mutode, mbondo*, and *muchaina* are the most popular varieties in Musikavanhu, where they are grown by 50 per cent or more of respondents. In Gokwe, *tsveta, mabeja, fumbati*, and *chitate* are the most common varieties, grown by 20 per cent or more of respondents. *Chidhomeni* is favoured because of its early maturity, *mutode* and *red tsveta* because of their suitability for beer and *white tsveta, mbondo, muchaina, mabeja, fumbati* and *chitate* are popular due to their white grains. *Muchaina, mutode* and *mabeja* were identified as having especially good storage qualities. Almost all the popular varieties are grown for successive years. *Mutode* is the most common variety in Musikavanhu and *tsveta* in Gokwe.

A large number of less common varieties are grown by many farmers in all the areas, but about half of the varieties in each area are grown by only a small proportion of farmers (often less than ten per cent of farmers).

Origins of varieties

Since the process of germplasm exchange is most probably an old and regular practice, distinctions between 'old', and 'new', 'traditional' and 'modern' varieties are hard to discern in most cases. In Musikavanhu respondents said that *mutode, mbondo, muchaina, chindindindi, dewe* and *ipwa* had been present in the area since the time of their grandparents at the turn of the century. *Chidhomeni*, a locally-adapted version of the officially-bred variety, Red Swazi (also known as *chibuku*), was said to have been introduced in Musikavanhu about 20 years ago by agricultural extension officers. In Gokwe, respondents who had immigrated into the area reported that they had brought seeds from their old villages.

Almost every farmer interviewed, when informed about the nature of the research, asked for varieties from other areas for testing on their own farm. This is indicative of farmers' interest and willingness to experiment with different varieties and points to the need for a more decentralized seed supply system, where farmers can choose varieties according to their own requirements.

The classification of varieties

Collective names are used for varieties that break the famine over the November to March period. *Mukadziusaenda* (in Musikavanhu) which means 'wife don't leave' (i.e. 'don't leave to go home to your parents because there is food to be eaten here') and *chinyanzwanzara* (in Gokwe), 'hunger conqueror' or 'hunger crusher', are generic terms which refer to several short-season varieties. *Chidhomeni* also ripens early, but respondents distinguish between it and *mukadziusaenda*, perhaps because *chidhomeni* is a more recent introduction.

Varieties may also be named according to their maturity period. In Siabuwa, short-season varieties of short to medium height (*lusiri* or *lusili*) are differentiated from long-season, giant varieties (Billing *et al.*, 1984). Alternatively, some varieties are classified according to gender roles associated with them. The pale-seeded or white varieties, *chidhomeni*, *mbondo* and *muchaina* are most frequently grown by women, while *mutode* is considered to be a 'man's crop'. This is because the latter variety requires much less work, since its tannin-rich grains are unattractive to birds.

Agronomic, gastronomic and morphological criteria

The criteria used by farmers to describe their varieties were grouped into three classes: agronomic, gastronomic and morphological. Agronomic descriptions included maturity period, soil and water requirements, tolerance of weeds, insects, pathogens and drought during the growing season as well as susceptibility to bird damage. Gastronomic descriptions included threshability, ease of winnowing, pounding and milling, good taste for beer and sadza (a traditional staple), colour of resulting food products, time required in cooking, keeping quality of the cooked grain, texture of endosperm and suitability for use in multiple food products, and storage quality. Morphological descriptions included grain and fodder yield, plant height, and tillering potential.

The frequency with which each type of description was used was analysed by matrix ranking. The most frequently employed criteria for distinguishing between varieties fell in the agronomic and gastronomic categories. In Musikavanhu, agronomic and gastronomic criteria were considered to be of equal importance while in Gokwe agronomic criteria were used more frequently than in Musikavanhu. At this stage it is not clear why gastronomic criteria were said to be more than twice as important in Musikavanhu than in Gokwe. However, in both areas, morphological criteria were least frequently referred to.

Although the culinary qualities and the taste of sorghum varieties are considered important gastronomic traits, in some areas early maturing varieties are tolerated despite their poor taste and poor storage qualities. Late maturing varieties are usually grown for their much-desired taste and cooking qualities, despite the higher labour inputs which they require in terms of weeding and scaring of birds. Higher prices are paid for beer made from a favourite variety, such as long-season *rongwe* (Billing *et al.*, 1984).

The varieties also differ in their resistance to storage pests. Weevils were identified as the main storage pest, but farmers did not express much concern about post-harvest losses of food grain. Some varieties such as *muchaina* (in Musikavanhu), *rongwe*, *mabeja* (in Gokwe) and *maila-tonga* (in Siabuwa) can

91

be stored for up to three years due to their hard, corneous endosperm. These varieties offer a measure of food security since they are available for consumption once the soft-endosperm varieties (such as *chidhomeni*, *mbondo*, *chivende*, and *chibuku*), which do not store well, have been consumed.

Similar findings have been reported for other small-scale farming systems. In the Philippines, gastronomic criteria were used eight times more frequently than morphological criteria when farmers were asked to distinguish between rice varieties (Nazarea-Sandoval, 1991). In the Dominican Republic, gastronomic criteria were considered so important that improved high yielding varieties of cassava were rejected on account of their taste (Box, 1984).

The influence of maturity period

Variation in maturity period plays an important role in the farming system of small-scale sorghum growers. Individual farmers generally grow two short-season, one medium and one long-season variety in different fields. Fifty percent of all varieties grown in Musikavanhu, Gokwe and Siabuwa are short-season. These ratios vary from area to area in relation to variations in rainfall. Mushita (1992) for example, found that in Chipinge, Zvishavane, Mutoko and Plumtree, 50 per cent of local varieties were medium-season.

Small-scale farmer variety selection

Farmers choose varieties for a number of reasons, many of which have not been given any consideration in formal sector sorghum-breeding programmes in Zimbabwe. The most important criteria identified in this study were, in order of importance, gastronomic criteria, early maturity, and agronomic criteria. Morphological criteria were of little importance to the farmers. In contrast, morphological criteria have received the bulk of breeders' attention (House, 1987; Obilana, 1988; ICRISAT, 1990). In no instance was yield identified by farmers as an important criterion in this study. The extremely low yields of less than one ton per hectare may partly account for this, but generally, small-scale farmers consider consistent, reliable and stable yields more important than high yields (Richards, 1985; Conway and Barbier, 1990).

Small-scale farmers are extremely vulnerable to the exigencies of external factors (such as off-farm incomes, environmental conditions and national policies) as well as internal factors (such as the demands of other crops in the farming system, labour availability and availability of animal draught power). Flexibility of farming operations is therefore an essential and fundamental component and is partly achieved by growing a number of varieties of different maturity period, each having different planting and harvesting times. Evidence from elsewhere indicated similar results. In Burundi, farmers rejected a new maize variety that gave up to a 40 per cent increase in yield because it did not fit in with the established agricultural routine (Haugerud and Collinson, 1990).

Small farmer conservation of sorghum genetic resources

In the survey area, while fields of cotton and other commercial crops were rigorously weeded, weedy and wild sorghum was allowed to mature in the fields of all varieties of cultivated sorghum. Although not articulated as such, this practice allows for a persistent, low-level gene exchange between cultivated varieties and weedy and wild sorghum. The process is controlled to some degree in that obvious 'off-types' are selected against by the farmer. The result is that genes which do not markedly affect the phenotype or which agree with the farmer's internalized concept of the 'archetype' of the variety, enter the genepool of the next season's seed stocks. These genes may confer qualities which have been selected for in the natural habitat, providing they do not have pleiotropic effects which affect the phenotype. Qualities such as disease resistance and drought tolerance may be incorporated into the cultivated crop in this manner. Evolutionary changes occurring in crop pathogens and crop pests, and changes in ecological conditions which are tracked by the weedy and wild sorghums, are thus incorporated in the cultivated crop.

Implications for the research system

In contrast to the interactive process of gene exchange which characterizes farmers' conservation activities, the reliance of scientists on conservation through the creation of *ex-situ* genebank collections isolates genetic materials from the very evolutionary processes which produced them. A more dynamic approach, which could incorporate aspects of small-scale farmer management of introgression, is needed for the conservation of sorghum genetic resources.

Haugerud and Collinson (1990) have pointed out that 'in Africa, the IARC's have spent more per head, hectare and tonne of food, with less to show for the effort than elsewhere'. The loss of financial returns has been partly a result of ecological and economic constraints (Baker and Siebert, 1984; Matlon and Spencer, 1984). However, the loss is also the result of a methodological mismatch which has its roots in the 'technological-package solve-all' approach adopted by green revolution agriculturalists. This approach creates an artificial divide between the farmers, their management of the farming system and the crop genetic resources. There is an urgent need for improved communication between farmers and breeders.

Maurya *et al.* (1988) developed a model of a participatory approach to breeding for improved varieties. Figure 2.1 adapts this to Zimbabwean conditions. The method conceptualizes small-scale farming systems as the central source of crop genetic diversity, which is complemented and supported by the national or regional genebank and breeding institutions. It aims to develop decentralized structures and its methodology incorporates four steps in a feedback-loop:

1. Selection of germplasm by breeders in consultation with the target farmers. The most popular varieties in an area (such as *chidhomeni, mutode, mbondo* and *muchaina* in Musikavanhu and *tsveta, mabeja, fumbati*, and *chitate* in Gokwe) should be selected as the primary building blocks of the programme.

2. Preliminary on-station selection based on criteria identified by farmers, and breeding for improved varieties with frequent feedback to farmers via agricultural extension workers.

3. Early release of a number of improved varieties for evaluation and selection by farmers under farmer conditions of management.

4. Co-ordination of farmer evaluations by agricultural scientists and breeders, and use of the farmer-selected material for subsequent trials.

The use of a participatory approach enables the efficiency and results of a breeding programme to be improved for two reasons. Firstly, a wider range of farmers can be targeted. Secondly, the utilization of farmers as selectors allows for rapid screening of a wide range of variables under conditions reflecting the constraints within existing agroecological systems (Ashby, 1987). It enables management constraints on breeding programmes to incorporate into research the numerous criteria considered important by farmers into research.

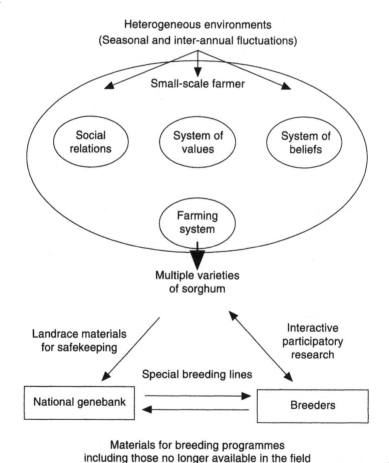

Figure 2.1: *Conceptual model of participatory breeding research*

A participatory breeding programme will also offer a form of local genetic resource conservation since the indigenous landraces will not be replaced by high-yielding introduced varieties. The need to improve local conservation of crop genetic resources and to preserve associated indigenous knowledge systems has recently been stressed at the international level (Keystone, 1991; SAREC, 1992). In Zimbabwe, both the national breeding programme and small-scale farmers stand to gain much from an approach which seeks to integrate the conservation of indigenous sorghum varieties and knowledge systems with breeding for improved material. Where breeding for improved material is integrated with conservation of indigenous genetic resources and knowledge systems, as in the participatory breeding methodology outlined above, farmers are empowered and remain the decision-makers in the selection processes.

Ex situ conservation by genebanks has been recognized and developed as the major focus for the conservation of crop genetic resources within policy frameworks. But the need for attention to be given to the neglected field of local crop conservation is extremely urgent. This requires a substantial shift in thinking about genetic conservation and cultural diversity and a meaningful commitment from the international scientific community concerned with crop genetic resources.

PART THREE

BUILDING LINKAGES

Introduction

Kate Wellard

A number of researchers and projects have been developing methodologies and institutional and organizational forms to strengthen the role of farmers in setting and prioritizing a research agenda. This section illustrates some of the different types of interventions which are being developed.

Some of the contributions are based in international and national research institutions, focusing on developing methodologies which can make them more responsive to the needs of farmers (Prain, Van Dorp and Rulkens). Others are more concerned with the researcher acting as a catalyst to enable local peoples to strengthen their organizational abilities, to use existing organizations in new ways to meet new goals (Tapia and Rosas, Taylor), and to facilitate their abilities to diagnose problems, plan and develop local decision-making (Mascarenas). Some contributions are concerned with promoting local cultural products, conserving biodiversity and promoting local crops which are often under threat (Tapia and Rosas, Mbewe). Wellard examines the roles NGOs have played in promoting local knowledge and enabling local people to present their needs to formal sector research and development.

Prain describes the experiences of international and national research programmes in Latin America and Asia in integrating local expertize in crop development with formal genetic resource research. Several innovatory research approaches are documented through which farmers have been able to participate in formal sector programmes as consultants (giving advice to researchers), as evaluators of genetic material, and as research curators who preserve local genetic materials. These initiatives involve local communities participating in programmes which are essentially researcher determined. The challenge raised is how to develop further initiatives in which scientists participate in programmes of locally-driven research.

Van Dorp and Rulkens show how researchers can develop methodologies which are sensitive to local perceptions in understanding local crop selection criteria, for the enhancement of formal sector research capacities. Researchers in Indonesia joined with local farmers to understand the processes and criteria of crop selection. Groups outside of farmers, such as traders and processors, with an important role in the marketing of crops, were found to have an important influence on the selection of varieties.

Tapia and Rosas describe how local knowledge of crops and soil in the Andean Highlands are being documented by researchers and used in formal research. Local seed fairs are important for farmers and are centres where they exchange material and knowledge about agricultural technologies. To encourage and facilitate the maintenance and conservation of genetic materials a group of NGO organizations are organizing varietal seed competitions at these fairs, which serve to revalidate and recognize the local knowledge and the innovations and crop development activities of farmers. This has met with considerable enthusiasm from the farming communities and it is planned to expand the programme.

The challenge of building on the knowledge and values of local people to sustain a meaningful development which is locally driven is explored by Mascarenas, in relation to the work of Myrada, a South Indian NGO. Using participatory rural appraisal methods and working with whole communities, the project has succeeded in bringing planning and decision-making down to the local level. It has even broken the monopoly of some highly-centralized government institutions.

Wellard documents the success of NGOs in linking between formal sector research and extension to promote and revalidate local knowledge. NGOs have encouraged researchers to work with farmers by providing logistic support, by organizing farmers into groups, and by helping to develop the capacity of farmers to diagnose problems for further investigation and research, and to plan, execute and evaluate their own experiments. By working alongside government researchers and extensionists and providing them with training, they aim to raise the understanding of the importance and relevance of local knowledge. NGOs also strive to increase the awareness among policy makers and consumers of the opportunities to be gained from developing local knowledge systems and the social, economic and environmental dangers of ignoring it. They are also supporting the initiatives of local organizations to lobby on behalf of their members.

The last two chapters also illustrate some of the roles NGOs are playing. Mbewe describes the work of Thusano Lefatsheng, a Botswanan NGO which is working with remote rural dwellers of the Kalahari desert, to document their knowledge and utilization of veld products, many of which are nearing extinction. They aim to reverse this trend and maintain the livelihoods of the remote area dwellers by promoting the sustainable utilization of veld products.

In a paper on black agriculture in South Africa, Taylor shows how communities have struggled to preserve their culture, traditions and indigenous knowledge systems in the face of hostile political systems and the 'modernization of agriculture'. He describes the work of CLIARD, in eliciting the needs of rural communities and farmers, and in researching with them their problems and implementing solutions. CLIARD is working to strengthen community institutions, building appropriate farmers' organizations and advocating the transformation of existing research and training structures. In this, it is concerned with the wider dimensions of the influence of dominant policies on community organization and agriculture.

The building of linkages between researchers and farmers arises from two objectives: the need to create more responsive research systems which cater for the needs of farmers, and the strengthening of the capacities of communities to undertake self-reliant development initiatives. In both cases the researcher or development worker faces problems which result from the marginalization of farming communities, local cropping systems and local crops. They are faced with the need to revalidate local knowledge systems, and to understand the socio-economic processes, political forces and cultural perspectives which result in marginalization of farming communities. This requires an understanding of the policy environment in which agricultural development is situated and the advocacy of policy reforms.

Mobilizing local expertise in plant genetic resources research

Gordon Prain*

In this paper, the author identifies three ways in which local R. & D. expertise can be integrated into genetic resources research: through participation of users as consultants, as evaulators of genetic material and as research curators of in situ *collections. These are illustrated by examples of research conducted by the International Potato Center (CIP) in Latin America and by The Users' Perspective within Agricultural Research and Development (UPWARD) programme in Asia. The paper concludes by considering how, instead of inviting users to join a schedule which is essentially researcher-determined, scientists might participate in locally-driven research.*

Global and Local Agricultural R. & D.

There is an urgent need to recognize the potential complementarity of global and local approaches to agricultural research and development (R. & D.). Given our lack of understanding of complex, resource poor systems and germ-plasm and breeding needs, it is logical to involve rural people as professional partners in research and development efforts. However, farmer participation in agricultural R. & D. goes beyond logic to issues of equity and social justice: farming families are the final users of agricultural technology and as such have most to gain (and to lose) from it.

Local-level R.& D. systems represent what Robert Chambers has called, 'the single largest knowledge resource not yet mobilized in the development enterprise' (Chambers 1983). Figure 3.1a illustrates early linkages between the global R.& D. system and 'peripheral' farming systems. One-way transfer of technology is developed largely through close ties with farming systems at the 'core'. A second opposite connection represents the extraction of products, especially crop ger-mplasm and, more recently, generalized information, in particular the understanding of systems and the registering of feedback in farming systems research (Horton and Prain 1989).

Mobilization of local R.& D. expertise in integrated genetic resources research can occur in three principal ways (Figure 3.1b): through participation of users as *consultants* (linkage A); as *evaluators* who receive new genetic material, evaluate it and provide information to researchers and disseminate the good material among the local population (linkages A,B,C); and as research *curators*

---

* Gordon Prain dedicates his paper to Mr Albert Navarro of Achipampa, San Juan de Jarpa, Junin, Peru. This farmer represents so many farming families in so many countries involved in the work reported in the paper.

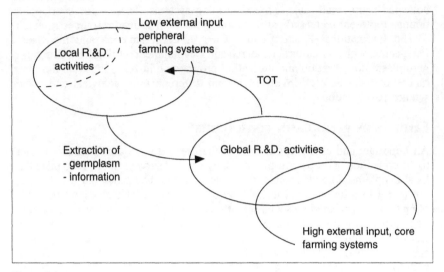

Figure 3.1a

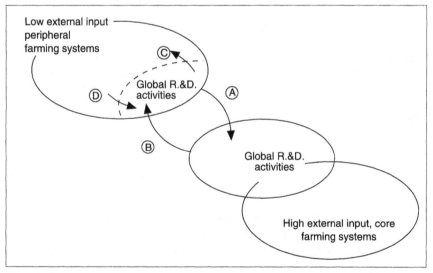

Figure 3.1b

Germplasm consultants   Ⓐ
Germplasm evaluators   Ⓐ   +   Ⓑ   +   Ⓒ
Researcher curators   Ⓐ   +   Ⓑ   +   Ⓒ   +   Ⓓ

Figure 3.1 a and b: *The need for complementarity between global and local genetic resources R. & D.*
(Adapted from Biggs, 1987).

of *in situ* collections, where users not only evaluate technologies coming from global R.& D. sources, but also receive, conserve and evaluate materials collected from small-scale, general farming systems. These three overlapping options represent increased participation by users in genetic resources research.

The International Potato Centre (CIP) in Latin America and the User's Perspective with Agricultural Research and Development (UPWARD), a rootcrop germplasm conservation and utilization programme in Asia, illustrate the three modes of local R.& D. involvement and differences in biological and social science participation.

Farmers as germplasm consultants

An important and urgent application of the use of farmer consultants is related to the collection of crop germ-plasm. Many varietal surveys tacitly recognize that local people are knowledgeable about the varieties they grow, but usually they fail to get to grips with the technically detailed nature of that knowledge. Too often they are prepared by social scientists alone and carried out by enumerators. The output turns out to be too general to be of much interest or use to breeders. Conversely, those scientists who are familiar with the technical aspects of germplasm, plant collectors and plant taxonomists in particular, rarely engage with local technical knowledge or, if they do, the knowledge is rarely processed. Remarkably little information is available on germplasm collections.

Interdisciplinary collection of germplasm and associated indigenous knowledge: from passport into biography

One way of conserving technical, socio-economic and cultural knowledge of cultivars is by joining users with plant collectors and social scientists in the systematic collection of both germ-plasm and its associated knowledge. There are at least three pools of information associated with plant germ-plasm collection:

o the genetic make-up of the plant;
o cultural knowledge about the plant; and
o cultural, socio-economic and ecological characterization of the plant's environment.

Current approaches to crop genetic resources mostly stress the first and third of these information pools (Prain *et al.*, 1992b). As far as we know, very few expeditions have given systematic attention to the second information pool: the indigenous knowledge associated with the germ-plasm. Information which is collected by conventional expeditions is referred to as 'passport data', regarded as the essential information for identification purposes: when and where the specimen was collected (longitude, latitude and altitude, country, province, nearest village, and distance from village).

Interdisciplinary collecting can turn the passport into a potted biography. Building on experiences in Latin America with expanded versions of the traditional 'passport data' form, a project has recently been initiated between the University of Cenderawasiah, Irian Jaya, UPWARD and CIP to explore

104

methodologies to collect and document sweet potato germ-plasm and associated indigenous knowledge in Western Java.

Using rapid rural appraisal tools a knowledge base is built up of the cultivars, the plot from which they were taken, and the system within which the plot is farmed. Group interviews are used to elicit consensual, comparative views on all the cultivars of a particular plot. These views are supplemented by individually-volunteered assessments during the collection itself.

This approach to collection of germ-plasm and associated local knowledge has its benefits and problems. Collecting germ-plasm is expensive and collecting indigenous knowledge increases the costs considerably. These extra costs are seen to be justified by the increased utility of 'known' material for other farmers and scientists. Although the detail of documentation is constrained by the utility of what is documented, it should not be totally determined by it. Use value either for genotypes or local knowledge cannot always be known. Biological and cultural diversity need conserving for an uncertain future.

'Memory Banking' of indigenous varieties and technologies

The interdisciplinary documentation of germ-plasm and associated indigenous knowledge described above is a means of greatly improving, broadening and enriching, but not fundamentally altering well-established germ-plasm collection practices which are still largely collector determined.

'Memory banking' proposes an alternative or parallel approach to current collection and conservation methods: it is the systematic documentation of indigenous practices of local farmers associated with traditional varieties of staple and supplementary crops. The approach stresses the dangers of cultural erosion as being equal to those of genetic erosion. The situation is particularly serious where the introduction of new varieties is displacing old varieties and with them indigenous practices and experiences (Sandoval, 1990). The conservation of genetic information is predicated on the fear that 'options for the future are being foreclosed by the erosion of one of the world's most important heritages, the genetic diversity of our crop plants and their wild relatives' (Plucknett *et al.*, 1987). The diversity of crop plants is largely a consequence of ecological conditions and human interventions. If we fail to document these, we decontextualize the genetic information in the sense that 'human and ecological forces which pushed for their selection will be largely ignored'.

The rhythm of memory banking is determined by the complexity of cultural knowledge rather than exigencies of plant collectors' schedules. User consultation is very intensive and the social researcher-user relation tends to take a more active role. In addition to RRA techniques, much more time-consuming methods are used: individual biography taking, verbatim recordings, 'mental mapping' (use of drawings by local people to get at the salient features of varieties stored in the memory) as well as extensive sampling of local cultivars for herbarium preservation. This type of documentation, the memory bank, has been completed for two communities of Bukidnon, Mindanao, the Philippines, with the bank currently located in the UPWARD offices. The next stage of the project is to explore methods for *in situ* gene banking and memory banking.

Farmers as germ-plasm evaluators

For thousands of years farmers have been identifying, evaluating and selecting naturally-occurring crosses in their own fields, especially in centres of plant domestication where crosses are possible with wild and weedy relatives (Harris, 1969). This 'slow-motion plant breeding', as it has been called, has often depended on the observational powers of women who historically have been most associated with seed selection and thus with noticing 'new varieties' which spontaneously appear in the field.

Modern plant breeding is certainly a lot faster than either natural selection or the local plant breeding described above but it can only be more successful if the materials it produces are adapted to particular environments and can be adopted by cultivators. The evidence so far for complex, low-input environments is not very good. Of more than fifty potato varieties officially released in Peru during the last thirty years, for example, only three or four are grown in over three-quarters of the area under modern varieties. One of the most important varieties in the north of the country was not even released officially. It was a clone rejected by breeders after trials, but kept, selected and multiplied by farmers (Franco and Schmidt, 1985).

Complementing modern plant breeding with local breeding and selection practices can sharpen the appropriateness of breeding goals and improve the adoption record. Complementarity can occur through identification of criteria used by farmers in selecting their crops which can improve breeding goals, or it can go a step further with farmers selecting the germ-plasm with which farmers are to work (Haugerud and Collinson, 1990).

Farmer-managed integrated seed programmes in Peru

A major problem facing small farmers in many developing countries is how to obtain good quality seed of desired varieties from an accessible location at the right time. Managers of seed programmes face difficulties in determining which varieties to include in expensive pathogen elimination, rapid multiplication and field multiplication, and how much of each variety to multiply (for example, see Crissman, 1989). In local R. & D., on the other hand, cultivar evaluation is integrated with seed multiplication and diffusion and can provide a potential model for a more appropriate, user-friendly seed programme. Figure 3.2 shows the interdependence between the testing of quite large amounts of new genetic material by farmers and the clean-up of a few farmer selections. The clean material is re-tested in a simpler set of trials with farmers to check relative performance under pathogen-free conditions. The results of evaluations are analysed for the range of adaptation shown and for the likely use. This analysis can be used to determine the amount of greenhouse space required for rapid multiplication.

Appropriate, farmer-managed methods of multiplication and distribution of seed could thus ensure the adequate supply of a variety to farmers who want it. Though such a scheme has never been implemented in its entirety, elements of it were explored as part of the Peruvian Seed Production and Distribution Project. Farmer participation in germ-plasm evaluation began by identifying twelve co-operators in six distinct agroecological zones (Prain et al., 1992a). A

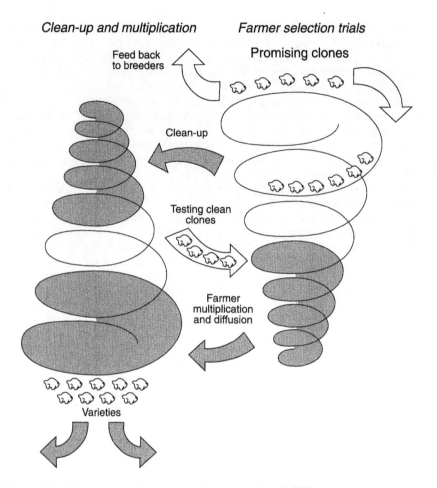

Figure 3.2: *User-managed variety selection and seed diffusion*

total of seventeen varieties were obtained from a number of breeding programmes and germ-plasm collections and were divided into two sets. The pair of farmer co-operators in each site were given one set each, so that all varieties were being tested in each zone, but each co-operator only had a maximum of ten varieties to evaluate. The same control variety, a very widely diffused, rustic modern variety, was used in all trials. An incomplete set of clones (because of lack of seed) was planted in the experimental station for comparison.

Production data were analysed by comparing average farmer production with experiment station production. The widely differing results obtained were likely to lead to quite different selections and eliminations by farmers and researchers. From the peasant potato producers six areas for evaluation and 39 separate criteria were identified. The importance of many of the criteria vary depending on whether production is destined for home or market use. Some of the criteria have apparently contradictory weights, reflecting alternative uses and benefits (earliness, for example, is desirable to secure the first fresh food of the season and higher market prices; lateness, on the other hand, is associated with higher yields and traditional, culinarily preferred indigenous varieties). Plant breeders cannot hope to deal with this kind of micro-adaptation and multiple evaluation criteria. What they can do is identify from the results of these trials certain crucial traits which they can deal with: for example, earliness, or pest resistance or storage characteristics. For the rest, they can leave the breeding populations they work with as heterogeneous as possible (within the limits of the likely target populations they are aiming for) and leave the farmers to identify the best-appreciated combinations of traits.

Two varieties evaluated in the Peruvian trials were clearly identified as appropriate for subsequent clean-up and farmer multiplication and diffusion, and another was introduced into the participative seed multiplication process. The process essentially involves two strategies which build on indigenous seed management practices, and the documented existence of seed flows linking communities through which farmers replenish their seed stocks (Prain and Uribe, 1990). The first strategy involves contacting communities, farmer groups or individual farmers in these key seed producing areas. The co-operators are loaned around 200kg of high quality seed from the rapid multiplication facilities of the seed programme. This seed is multiplied in plots which offer reasonable isolation from other potato fields. The loan is returned (for redistribution to other users) and distribution of the clean seed occurs among the existing kin or neighbours based channels. The second strategy is based on the farmer practice of obtaining very small quantities of good seed from markets or other outlets and then slowly multiplying the seed over several generations until the required quantity is produced. Small lots of the high quality seed are sold to farmers via the extension service, rural development project or local markets, and multiplied. Follow-up studies found that three-quarters of farmers who obtained seed in this way kept all or most of the harvest for re-multiplication as seeds (Prain and Scheidegger, 1988).

Farmers as research curators

So far we are still talking of *integrating* the user's perspective into genetic resources research, of farmers participating briefly in our show. What are the chances of going one step further, of us participating in an on-going, local R.& D. process?

A possible entry point for a radical transformation of current practice can be found in the ideas surrounding *in situ* or communal germ-plasm conservation. One solution to the problem of decontextualized germ-plasm and the rapid erosion of germ-plasm from *ex situ* collections (Fowler and Mooney, 1990) is for rural people, many of whom already practice a discrete form of biodiversity conservation in their tree lots and backyard gardens, to take on the task themselves, with support from government and NGO agencies.

In situ conservation of rootcrop diversity in the Philippines

An UPWARD project in the Philippines is beginning to explore some of the conceptual issues and methodological problems of *in situ* conservation in two sites in Bukidnon, Mindanao which offer the possibility of comparing different geographical settings and different types of potential curatorship.

One type of curator collective consists of an informal grouping of about 19 migrant women in Maambong, Libona who are engaged in a moderately commercialized production of sweet potato and other rootcrops as well as maize, the principal staple. Vegetables are grown as a cash crop. The women have formed themselves into an association known as Industrious Mothers and a senior member has donated a piece of land for the genebank. Each woman has been allotted one row of the genebank in which to plant the varieties of sweet potato, taro, cassava and yam which they have obtained. The other group was to consist of the formal, male political authority of the indigenous community of Dalwangan. Unlike at the first site, varieties of sweet potato, taro and cassava were pooled and jointly planted.

Asking agricultural technology users to take centre stage after years of being treated as recipients of technological largesse at worst, and information pools at best, represents a radical shift of approach. To get things going requires a lot of support, motivation and encouragement from researcher partners and even some material incentive. Incentives were much less important for the women's group, where considerable friendship and trust had been built up during an earlier memory banking project. Nevertheless, it was agreed to support the installation of a small water tank to be used as emergency irrigation for the genebank should there be danger of losing the collection. With the second group, no previous contacts existed, the researchers lacked established social relations with the local people, and there was an expectation on the part of the authorities that a more tangible benefit was needed to stimulate initial interest than the bringing together and preservation of rootcrop germ-plasm. It was agreed that some goats would be purchased and materials for a goat-house provided.

For both groups Certificates of Membership in their association were prepared and in the case of the women's association, it was agreed to present prizes for the greatest diversity maintained in an individual plot within the genebank. At harvest, visits will be arranged for members of both groups to the other site in order to stimulate discussion and exchange experiencesw.

Though it is too early to draw conclusions, the experience so far with the 'Industrious Mothers' group is extremely positive. Of course, the essence of community genebanking is its long-term viability, and a proper evaluation will only be possible after several seasons. Nevertheless, the belief in the importance of conservation is already fixed within the group. Though there are certainly several problematic elements already apparent in the second site, there are also some extremely interesting developments and the owner of the land on which the genebank is based has already begun to explore the potential use of the genebank accessions.

The hope is that the local curators of the genebank will become the local evaluation specialists, a role which would take on more institutional significance and hence offer more prestige, as with healer or document writer, than in the past. The question remains, however, as to whether the treatment of the collection as active might undermine the conservation function. To help balance these two functions, it will be important to support the notion that possessing a large number of different tasting, different performing varieties is a source of prestige within the community and is an important resource in times of calamity and for future generations. These dynamics remain to be explored and tested over the next seasons.

Moving towards locally-determined research

The cases documented here show the detailed knowledge and wide-ranging practices which rural people utilize in the conservation, evaluation and exploitation of plant genetic diversity. The failure of the global R&D system to recognize that expertise has led to vast, anonymous and erosion-prone collections of crop germ-plasm across the world which have been of limited use to other farmers or crop scientists. It has also meant that conventional plant breeding and seed programmes have had very limited impact on the complex, resource-poor farming systems in which the majority of the world's farming families must make their living.

This vast pool of untapped, local R&D expertise must be linked up with global R&D activities. Local R&D practitioners have the skills and knowledge to evaluate experimentally unfamiliar germ-plasm as active partners in breeding research and then to multiply and disseminate clean seed of those varieties within user-friendly seed programmes. Biological scientists should be encouraged to become close partners with local evaluators in order to optimize global R&D understanding of local evaluations.

Fuller recognition and performance of local R&D as a legitimate research mode can be achieved more effectively through community curatorship of plant species and varieties for conservation, research and effective utilization. With this radical shift in agricultural R&D's centre of gravity, participation takes on a different meaning. Instead of inviting users to participate in a schedule which is essentially researcher-determined, social and botanical scientists have the chance to participate in a process which is genuinely driven from the local point of view.

Seed fairs in the Andes : a strategy for local conservation of plant genetic resources

Mario E. Tapia and Alcides Rosas[*]

This paper provides interesting insights into local knowledge of soils and crops in the Andean Highlands, the value of which is now being recognized and documented. It describes local seed fairs where material and information about agricultural technologies are exchanged between several communities in a region. A group of researchers and NGOs are encouraging peasant farmers to maintain their seed diversity and maintain in situ conservation of genetic material by organizing competitions between members at these fairs. Ways of promoting this innovative programme on a regional scale are now being investigated.

In situ conservation, or the maintenance of genetic resources in natural environments, has generally been considered by researchers to be an untenable strategy for crop species. Evidence from various parts of the world, however, shows that conservation has been maintained over a long period of time. This paper discusses methods being used to promote conservation in the Andean highlands (King, 1988).

At the time of the Spanish invasion in the sixteenth century, the inhabitants of the Inca empire and other kingdoms in the Andes were cultivating 70 separate crop species (Cook, 1925). Most of these are still under cultivation in traditional communities and in some cases in isolated home gardens. Highly diversified agroecological conditions and Andean traditional farming systems mean that a continuous flow and selection of seeds is necessary to maintain diversity and reduce risk in agricultural production. Thus it appears that a strong system of interchange of reproductive material has been needed. Local and regional seed fairs have played an important role in this process.

Characteristics of Andean agriculture

Agricultural production systems in the Andes vary significantly with altitude and latitude. Locations from Venezuela to Ecuador are generally more humid than those farther south, while eastern slopes receive more precipitation than the valleys facing the Pacific Ocean, especially in the Central Andes in Peru.

---

[*] Mario Tapia and Alcides Rosas dedicate their paper to the farmers from the communities of Chamis and La Escañada, from Cajamarca, Peru. They participated during the last three years in the organization of the seed fairs and gave the authors all their experiences in the conservation of the genetic material.

Altitude has a profound effect on crop adaptation. At higher altitudes the incidence of low temperatures during the growing period defines the range of various crops and varieties. Frosts are frequent at altitudes over 3000m, and particularly so below the eight degrees southern latitude from Peru to Bolivia, where a frost resistant potato was has been domesticated (bitter potato or 'papa amarga': *S. juzepzukii, S. curtilobum*).

In the valleys (Quechua region) crops grown include maize and beans, together with other, so-called minor, crops such as amaranths, quinoa and cucurbitaces. At this altitude and down to the 'yunga' area (1500 to 2500 m), native fruits are cultivated, especially in home gardens, in a very intensive system of production mixed with root species such as arracacha, yacon and mauka. In mid-altitudes, including slopes (2800 to 3800m) and the high plateau, potatoes and others tubers, oca, olluco, mashua, are rotated with native grains, quinoa and lupine, or with introduced crops, barley and faba bean. At higher altitudes (Puna region, 4000 to 4500m), other crops such as resistant potato, the root, maca, and some kañiwa (native Chenopod grain), forage barley and oats are the only species to be cultivated.

Local crop diversity

Food species domesticated in the Andes include grains, tubers, roots, fruits and species which are adapted to difficult environments and soil conditions and provide a varied diet. Many of these have been cultivated since ancient times or originate from then. Thirty such species from the High Andean region are shown in Table 3.1.

Local classification of soils

Soil classification and soil maps of mountain regions, particularly the international soil classification systems presently in use, are very often inadequate and more attention needs to be paid to producing maps which can be used in land-use planning. Relying on the often highly developed and practical indigenous classification system can bring better results (Mateo and Tapia, 1990).

One well-documented local soil classification system is that of the Quechua-speaking communities located in the highlands of Cusco Department, Peru (Rozas, 1985). Soils are characterized by means of indicator plants and according to topographic, climatic and other characteristics. Peasants have thus developed an applied system which is based on experience and observations and suits their needs well. A few examples with names in Quechua, the native language of the Inca Empire, are shown in Table 3.2. The system also provides very precise and detailed classification by irrigation, humidity, texture and soil compaction. It is widely used in allocating land to community members according to their social status, marital status and age.

Socio-economic traditions

In addition to the different agroecological conditions, distinct patterns can be discerned in social status and land tenure. Agricultural production systems depend on the interaction between ecological, social and economic factors.

Table 3.1: Food species originating from/cultivated since ancient times in the High Andean Region

Common name	Scientific name	Botanic family	Altitude for optimal growth (m)
Tubers			
Potato	Solanum andigenum	Solanaceae	1000-3900
Bitter potato	Solanum juzepczukii	Solanaceae	3900-4200
Oca	Oxalis tuberosa	Oxalidaceae	1000-4000
Olluco	Ulucus tuberosus	Baselaceae	1000-4000
Maswha	Tropaeolum tunerosum	Tropaeolaceae	1000-4000
Roots			
Arrachacha	Arracacia xantorrhia	Umbelifereaea	1000-3000
Achira	Canna edulis	Cannaceae	1000-2500
Ajipa	Pchyrhizus tuberosus	Leguminoseae	1000-2000
Yacon	Polymnia sonchifolia	Compositeae	1000-2500
Chago	Mirabilis expansa	Nyctaginaceae	1000-2500
Sweet Potato	Ipomea bataras	Convolvulaceae	0-2800
Maca	Lepidium meyeneii	Crucifereae	3900-4100
Grains			
Maize	Zea mays	Granineae	0-3000
Quinoa	Chenopodium quinoa	Chenopodiaceae	0-3900
Kaniwa	Chenopodium pallidicaule	Chenopodiaceae	3200-4100
Kichiwa	Amaranthus caudatus	Amaranthaceae	0-3000
Legumes			
Tarwi	Lupinus mutabili	Leguminoseae	500-3800
Beans	Phaseolus vulgaris	Leguminoseae	100-3500
Lima bean	Phaseolus lunatus	Leguminoseae	0-2500
Pajuro	Erythrina edulis	Leguminoseae	500-2700
Cucumber			
Zapallo	Cucurbita maxima	Cucurbitaceae	500-2800
Caygua	Ciclanthera pedata	Cucurbitaceae	100-2500
Fruits			
Pepino	Solanum muricatum	Solanaceae	800-2500
Tomate de arbol	Cyphomandra betacea	Solanaceae	500-2700
Lucuma			
Chirimoya	Lucuma aborata	Sapotaceae	0-2500
Aguaymanto	Annona cherimola		500-2000
Tumbo	Physalis peruviana	Solanaceae	500-3000
Tintin	Passiflora mollisima	Passifloraceae	1500-3000
	Passiflora pinnatispula	Passifloraceae	2500-3800
Species			
Chile, aji	Caspicum annum	Solanaceae	200-2000

Table 3.2: Local soils and land use classification in Cusco Department, Peru

Local name	Scientific name	Description
Classification by indicator plant		
Llapo pasto	Muhelmbergiha peruviana (a Gramineae)	Growth indicates very poor soils; yields will be lows.
Layo	Trifolium peruvianum	Grows only in compacted and stonly soils; soil preparation will be an important limitation.
Pilli pilli	Hypochoeris sp	Grows well in wet and swampy soils; not appropriate for farming
Salvia	Slavia officinalis	Indicates high quality and fertile soils well protected from frost.
Classification by location		
Puna allpa		Lands located at Puna altitudes and very prone to night frosts
Yunga allpa		Warmer lands, not prone to frost.
Classification by climate		
Chiri allpa		Land in cold zones appropriate for dry farming, where the layme system (rotational cropping) is practiced
Qoni allpa		Land intemperate zones used for annual cropping, normally irrigated, with no fallow period.
Classification by topography		
Qayqo allpa		Lands found in deep valleys
Pampa allpa		Flatlands
Qhara allpa		Slopes with thin soils
Moqo allpa		Hilly lands
Phukru allpa		Steep slopes

Source: Rozas, 1985

In the high Andes (Ecuador, Peru and Bolivia), the traditional social organizations of peasants or 'comunidades indígenas' (indigenous communities) are formed by descendants of the pre-Hispanic, extended family units. In Peru alone they number more than 4000 and maintain many traditional technologies for agricultural production. Whilst there are also significant numbers of small private farms (and, in Ecuador, large farms) as well as new associate organizations, the indigenous communities form the great majority of local organizations.

Native communities in the Andes have preserved plant genetic resources despite neglect, and even scorn, by much of the society around them. Even today, at local markets or fairs, women in distinctive hats and homespun jackets sit behind sacks of glowing grains, baskets with beans of every colour, and bowls containing traditional fruits. At their feet are piles of strangely shaped tubers, red, yellow or purple coloured (NRC, 1989).

Seed fairs

Over many centuries, people from native communities have congregated at certain places on established days during the year to exchange goods from different valleys or regions, a practice which continues today. A vast cropping field can turn into a lively camp for a few days.

A distinction should be made between the weekly fairs which form part of the ordinary process of marketing domestic goods, and these regional fairs which are usually held only once a year, often during a religious festival, where communities belonging to different agroecological zones congregate. Dates coincide with the end of the harvest so that peasants can buy provisions of food, textiles, tools etc, and at the same time have the opportunity to exchange different seeds or replace those lost due to climatic or economic factors.

Without doubt these regional fairs act not only as commercial markets, but are also considered to be an opportunity to exchange material and knowledge about agricultural technologies between several organized communities in a region. Some of the traditional Andean fairs where exchange of plants and seeds takes place are shown in Table 3.3.

Recognizing their important function, a group of local researchers and NGOs suggested that the seed fairs could provide an opportunity to encourage peasants to maintain their seed diversity and to stimulate a practice already being followed by many peasants, *in situ* conservation of genetic material (landraces). This is particularly important now that traditional systems appear to be weakened by the economic crisis affecting the region. The idea was to organize a competition between community members to give recognition to seed diversity and related agronomic knowledge held by the peasants.

Methodology

Twelve NGOs working in the area participated with staff from three local universities and the agricultural research institute, Instituto Nacional de Investigación Agropecuaria y Agroindustrial (INIAA), in organizing these competitions. Several steps were followed:

Agroecological zonation. Most of the crops are related to a specific agroecological zone and so subregions have been defined to facilitate and focus the work. The Peruvian Andes cover more than 32 million hectares and given poorly developed transportation facilities, two different regions were chosen initially. These were northern, more humid region and the semi arid southern part of Peru.

Table 3.3: Traditional Andean fairs

Location	Date	Region/theme
Regional fairs		
Copacabana/Bolivia	6 August	High plateau of Peru and Bolivia
Pucaru/Peru	19 July	High plateau, Peru
Tiobamba/ Cusco/ Peru	15 August	Southern Peru
Theme fairs		
Cusco, Corpus Cristi/ Peru	June	Fruits
Cusco/Peru	24 December	Medicinal herbs

Determination of the gene-centres. In each region, locations were selected where the maximum diversity of a given crop is found. Factors such as the presence of a rural development project were also considered. Some centres of diversity are very well known, like the areas where bitter potatoes are cultivated in Puno, or the Puna agroecological zone, in south Peru and Bolivia, the only places where kañiwa can be found. Many varieties of quinua are grown around Lake Titicaca and in valleys such as the Urubamba, the Mantaro and the hillsides of Cajamarca. Tubers like oca, olluco and mashua are typically found in the Suni agroecological zones, in rotation with the potato crop. The western slopes in Peru such as Cuyo, together with the most humid areas at altitudes of 3600 to 3900 metres, are recognized as centres of tuber concentration. These are considered to be priority areas in the *in situ* conservation process.

Formal invitations to the communities. Two to three months before the fairs a letter of invitation was sent to the president of each community indicating the objectives, date, place and rules for the crop diversity competitions.

Field registration and evaluation procedures. On the day of the seed fair competition, each peasant or group of peasants were registered and interviewed to establish:

o species presented;
o name of the varieties or local landraces;
o morphological and sanitation characteristics of the material; and
o agronomic information and knowledge.

Prize nomination. The juries were composed of two technicians with experience of genetics and germ-plasm, and two peasants known in the zone to be good agriculturalists. Depending on the number of participants, prizes related to a single crop or to a group of crops such as potato or tubers. The winners were the three to six participants who presented the most diverse crop material. Prizes included agricultural tools and sacks of seeds and a certificate of participation. Winners were also accorded an important degree of recognition by all community members.

Table 3.4: Winners of the seed fair in Cajamarca (La Encañada), June 1990

Name of peasant	Community	No. of crops	No. of potato landraces	Local accessions
Francisco Aguilar	Mangle	8	7	40
Miguel Riquelme	Usnio	8	6	29
Rosa Abanto	La Torre	7	11	23
Mothers Club	Usnio	9	3	16
Francisco Saldaña	Encañada	6	5	16
Pablo Gallardo	Aguamala	2	-	23

Outcome of the seed fairs

The methodology involves not only the evaluation of genetic material present in one area but also the peasants' knowledge and the agronomic requirements for production of different varieties. Thus the process becomes a means of diagnosing agricultural problems and identification of opportunities and can be used to define the research and extension activities within a rural development programme. Individual peasants or communities are selected as 'conservation units' and a follow-up process is started. In several locations women have been chosen to be responsible for seed selection, as they know how to cultivate different varieties and use the material for food preparation, and they maintain facilities for seed conservation.

Over thirty fairs have been organized so far. Some of the peasant participants have been found to maintain more than 20 different species in their 'chacras' (plots). The results of one fair held in Cajamarca, La Encanada, are given in Table 3.4. Peasants are growing up to 48 different varieties of potatoes, tewlve varieties of beans, or eight of maize and may be considered 'germ-plasm conservationists'.

Special workshops have been held with peasant conservationists to discuss issues such as genetic conservation value and agronomic and economic evaluation. Support is being sought for conservationists, in particular agronomic support. Ways of maintaining a programme of seed conservation and production on a regional scale are currently being investigated.

Farmer crop-selection criteria and genebank collections in Indonesia

Marianne van Dorp and Ton Rulkens[*]

The paper describes research carried out on the Indonesian islands of Lombok and Sumbawa, in which local varieties of soybean, maize, cassava and sweet potato were collected for inclusion into the national genebank and farmers' knowledge of these local varieties was investigated. A multidisciplinary research team carried out informal interviews with individual male and female farmers, farmer groups, and other key persons. Data were collected on agronomic and product quality aspects of varieties, and changing patterns of variety selection through time. The paper shows the complex interaction of factors that influence the selection of crop varieties. It points to the importance of investigating local perceptions and developing multidisciplinary research approaches to gain an understanding of the processes of farmer crop selection.

Introduction

The collection of data on farmers' knowledge of varieties and landraces can provide useful information which may be used to classify germ-plasm stored in genebanks. This information reveals the selection criteria used by farmers. These could be integrated into breeding programmes, guiding them towards producing more relevant varieties. Within the Malang Research Institute for Food Crops (MARIF), which houses the Indonesian national genebank for non-rice food crops, a research project was designed to study farmers' knowledge of local varieties/landraces of food crops and to collect genetic materials for the Institute's four major mandate crops: soybean, maize, cassava and sweet potato. The little-researched province of West Nusa Tenggara, consisting of the islands of Lombok and Sumbawa, was chosen as the study area.

This chapter describes the research approaches and methodology used by the project; the criteria found to influence farmers' selection of varieties of soybean, maize, cassava and sweet potato; the complexity of factors affecting the selection process; and the implications for developing multidisciplinary research which responds to the conditions and needs of farmers.

---

[*] Marianne van Dorp and Ton Rulkens dedicate their paper to the farmers of Nusa Tengara who shared their knowledge with them.

Methodology

Earlier research work carried out at MARIF revealed that farmers' decisions to adopt or reject particular varieties were influenced by various agronomic and product-quality criteria (van Dorp and Utomo, 1989; van Dorp and Utomo, 1990; Ruthens *et al.* 1989/1990). From these criteria a framework for the collection of data through informal interviews was drawn up (Table 3.5). More attention was paid to the aspects of acceptability of varieties than to the availability of planting materials. A historical dimension was added to the checklist to understand farmers' reasons for shifting to other varieties.

Research was based on rapid rural appraisal techniques (RRA), combining a study of existing documentation with informal interviewing. Field observations were important in both complementing and verifying information obtained. Interviews were held with individual male and female farmers and farmer groups

Table 3.5: Framework for data collection on farmer varietal selection

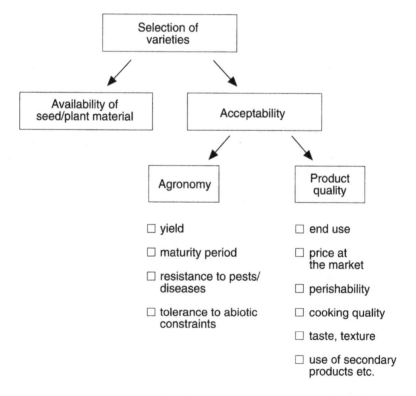

Note: The research project was among other things meant to 'test' the framework. The framework was set up by the authors to guide the research process.

as well as with key persons such as traders, retailers and agro-processors. Secondary data were used to identify the main production areas for soybean, maize, cassava and sweet potato. Villages within these areas which are major producers were selected for study since it was felt that farmers there would have a large breadth of knowledge of the varieties, while the collection of genetic material would be relatively easy.

The research team consisted of between five and eight researchers, the actual composition of the team varying with the crop under study. An interdisciplinary approach was pursued with all members of the team using the complete checklist for interviewing. Disciplines represented included genetic resources, agronomy, plant breeding, post harvest handling, food processing and human nutrition and team members were drawn from the research institute and the agricultural extension service. Researchers and extensionists learnt from each other, using the other's information to understand the farmer's situation better.

Soybean selection

Soybean is the most important leguminous food crop of Indonesia, and provides cheap protein-rich food products. Most soybean is consumed as *tahu* (tofu) or *tempeh* (fermented soybean cake). Since the 1970s domestic demand for soybean has outgrown production and increasing quantities of soybeans have been imported. For this reason, increased soybean cultivation is being encouraged in areas outside the traditional soybean growing areas of Java and Bali. The province of West Nusa Tenggara is an important new production area of soybean, and is now the third largest soybean producing province of Indonesia, accounting for ten per cent of the national output. Most of this production is transported to the densely populated island of Java. Minor quantities are processed into *tahu* and *tempeh* within the province.

Twenty-one villages were visited and around 200 interviews conducted. Over one hundred samples of genetic material were collected and stored in MARIF's genebank. Their main characteristics are described in Table 3.6.

The government is also attempting to increase yields through soybean intensification programmes. The use of certified seed of the improved variety, *Wilis,* is an important part of this. However, certified seeds of improved soybean varieties are often not available and are more expensive than seeds bought at markets or from other farmers.

The majority of farmers plant non-certified seeds of landraces or varietally-impure improved varieties. For soybean, an informal seed supply system exists, in which soybean seeds circulate throughout the year between different agro-ecological areas:

o Non-irrigated uplands. Soybean is grown as a sole crop in the rainy season, harvested from March-April;

o Rainfed rice lands. Soybeans are planted after a single rice crop in the period April-May and harvested from July-August;

o Irrigated rice lands. Soybeans are planted after two consecutive rice crops in July-August, harvested from October-November.

Table 3.6: Number of accessions and most important characteristics of landraces from West Nusa Tenggara.

Crop	Number of accessions	Characteristics reported by farmers in different landraces
Soybean	59	resistance to insect pests, tolerance to waterlogging, tolerance to intermittent drought, small seeds, mixed seed colours, slightly bitter taste
Maize	34	white seed colour, short maturity, in one variety combined with relatively high yield, good storability, good consumption characteristics
Cassava	7	sweet, good storability, extendible harvesting time, good consumption characteristics, leaves suitable for animal feed
Sweet potato	27	good consumption characteristics, good storability, leaves (very) suitable for human consumption, long maturing, extendible harvesting time

This system guarantees viability of soybean seed, but not varietal purity. The system implies that only varieties which are sufficiently well adapted to all three seasons/agro-ecological conditions are likely to be adopted by the farmers. The improved variety *Wilis* is rejected by many upland farmers. They noticed a high degree of seed mottling (probably caused by soybean stunt virus) when *Wilis* is grown on rainfed uplands during the rainy season. Mottled seeds are considered not suitable for sowing-seed and have to be sold as consumption seed, which fetches a much lower price.

Many farmers who grow soybean on rainfed rice lands prefer landraces to modern varieties because landraces are said to perform much better under the waterlogged conditions which prevail in many parts of these ricelands. Modern varieties can be grown on these soils only after digging furrow drains which requires a substantial labour input. Resistance to pests was defined as an attribute of several old landraces. The variety *Godek* (or 'monkey', referring to its densely pubescent stems, pods and leaves), a very heterogeneous landrace from Lombok, is said to have some resistance to the pod sucking bug *Riptortus linearis*.

In terms of quality, large-seed modern varieties (100 seed weight: 9 to 12 grams) are generally regarded as superior to the small-seeded landraces (100 seed weight: 6 to 8 grams). Landraces often have a high proportion of green and brown seeds, whereas modern varieties are homogeneously yellow. In *tempeh* processing, small seeds with mixed colours are not appreciated. Peeling small seeds takes more time and peeling of mixed coloured seeds should be more thorough because good quality *tempeh* should contain homogeneously yellow soybeans.

Maize selection

Maize is mainly grown on non-irrigated uplands, often intercropped with cassava, soybean or other legumes. In the past, maize was a rice-substituting subsistence crop but now most maize is either sold as dry kernels to the chicken-feed industry or retailed as young roasted cobs along the roadside and in urban centres.

The chicken feed industries, located in the urban centre of Mataram, West Lombok, require yellow seeded maize. In response to this demand, most farmers in Lombok have shifted from the low-yielding white seeded local varieties to yellow seeded high-yielding hybrids. Certified hybrid seed is considered to be too expensive by most farmers (3500 Rupiah/kg). Farmers prefer to purchase non-certified hybrid seed, which is available at 500 Rupiah/kg or less, or they retain their own seed. They are aware of the decrease in yields after each consecutive generation of retained seed, but consider that even after several generations the yield of hybrid-derived varieties is still higher than the yield of local varieties.

The local white seeded varieties are still preferred for subsistence use and for selling as young maize. Maize grits are mixed with rice in larger or smaller amounts, depending on the prevailing economic situation. Hybrid maize is considered inferior for human consumption because of poor taste and a hard texture. The white colour of the local maize also makes it a more attractive rice substitute.

Farmers who grow predominantly hybrid maize often plant a small part of the field with local maize. Local maize matures earlier (75 to 90 days compared to 100 to 110 days for the hybrids) and can fill in the pre-harvest gap in food availability. Local varieties are preferred in the drought-prone soils and climate of southern Lombok, where the long maturity of hybrid maize is considered too risky. In Sumbawa, local maize varieties still dominate. Yellow seeded maize is not in demand here and maize is predominantly used for human consumption. There is a wide diversity of maize varieties in the eastern part of the island, with differences in seed colour (white, yellow, red and mixed), seed type (flint, dent, glutinous maize) and maturity period (from 70 days to 110 days). In many parts of sparsely-populated and densely-forested Sumbawa, wild pigs are the major pest for maize and other food crops. Fields bordering forests have to be watched continuously during the night. In some areas an early maturing glutinous maize variety is planted in soybean fields as a 'trap crop': whenever maize is present pigs will not touch the soybeans. Wild pigs feeding on maize make a lot of noise and thus alert the watchmen. According to the farmers, wild pigs also have clear preferences for certain varieties of maize and soybean.

Cassava selection

The northwestern part of Lombok is the only cassava-producing area on the island. The local name for cassava, *Ambon Jawe* ('Java root tuber') points to a Javanese origin. While a wide diversity of cassava varieties are grown in Java, in Lombok not more than seven clones could be collected after extensive searching. Sweet and bitter varieties are grown in Java, but Lombok farmers only cultivate sweet cassava. Cassava starch factories process most of the bitter cassava in Java, but there is none in Lombok.

The dominant cassava clone in northwest Lombok is *Kuning* ('yellow', referring to the yellow root pulp). This clone was introduced about ten years ago, probably via farmer-to-farmer contacts, and rapidly became the main cassava variety. Most of the harvested roots are transported to the urban centre of Mataram where the roots are processed into several snack foods, the most important of which is *tape*, a sweet and slightly alcoholic product made by fermenting boiled cassava roots. The cassava processors prefer roots of *Kuning* because of the attractive yellow colour and the dry texture of the *tape*, probably caused by the high dry-matter content of roots as revealed in extensive screening of germ-plasm at MARIF. The strong preference of processors for *Kuning* means that roots of other clones are rejected by cassava traders, and this in turn influences farmers' choice. Other characteristics of *Kuning* identified by farmers as favourable include:

o good yields;
o can be harvested fairly early: from 6 months onwards;
o retains the leaves even in the dry season when other clones shed
  their leaves (probably related to resistance to Tetranychid mites);
o crop residues can be used as cattle feed.

However, farmers still grow traditional clones in limited amounts for home consumption. They argue that *Kuning* gives an uncomfortable feeling in the stomach when eaten in large quantities.

Livestock is an essential part of the farming system in northwestern Lombok. In the dry season, green cattle-feed is scarce and cassava leaves and stalks become an important input. This influences the choice of cassava varieties grown by farmers. The clone *Banyuwangi* has good consumption and agronomic characteristics but was quickly rejected by the farmers because its crop residues had an intoxicating effect on cattle.

Cassava has been an important subsistence crop in northwestern Lombok up until the 1980s. To safeguard food availability throughout the year, several clones were grown which could be harvested at different times. This included short-duration varieties such as *Datu*, which could be harvested from 4 to 5 months after planting, and long-duration varieties such as *Besi*, requiring a minimum of 11 months before harvesting. In times of abundant cassava production the maturation of *Besi* could be delayed by cutting down the shoots at the start of the rainy season. The roots could then be harvested in the following dry season. *Besi* is also known for good storability of harvested roots, which can last for one week compared to 3 to 4 days for other clones.

Sweet potato selection

Central Lombok is the main centre of sweet potato production in the province. A total of 20 different clones were found in two villages. As with cassava, sweet potato production has recently been reoriented from household subsistence to market production. Most roots are transported to the urban centre of Mataram where the roots are consumed as snack foods. *Sokan*, a group of three fairly similar clones, is presently dominant. Boiled roots of *Sokan* are esteemed for their consumption characteristics, including a sweet taste and a dry texture. Other positive characteristics are the good storability of roots, high yields and early maturity.

During the period when sweet potato production was primarily for subsistence, *Kanjon* was a very popular clone. This was attributed to its large roots with excellent storability and resistance to attack by sweet potato weevils (*Cylas formicarius*). Boiled roots of *Kanjon* have a firm texture and initially have a non-sweet taste which becomes sweeter during storage. Because *Kanjon* is a late-maturing variety (it can be harvested five months after planting) it was planted together with varieties which can be harvested earlier, mainly *Sokan*.

The varieties *Gule* ('sugar') and *Marak* ('red', referring to the colour of the root skin) produce roots which are extremely sweet and wet-textured. They are preferred for some ceremonial dishes and therefore fetch a very high price when sold in the local markets. However, these clones are usually grown only for home consumption because they have very low yields and have roots with extremely poor storability.

The varieties *Bedug* (a large drum, referring to the round and extremely large leaves) and *Kangkung* ('water spinach') are grown as perennial climbers supported by trellis. They are specifically used as a green vegetable and prepared in the same way as water spinach (*Ipomoea reptans*). Both varieties have non-sweet roots with a wet texture which are rarely consumed.

Implications for research of local selection and use

The research revealed that farmers' criteria for selection of varieties involves a complex interaction of agronomic, consumer preference, and socio-economic and changing market factors. The multidisciplinary composition of the team helped in gaining a better understanding of these complexities. The technical scientists from post-harvest and food technology were also important additions to the team, as product-quality aspects often turned out to be crucial in the choice of varieties by the farmers.

An awareness of the division of labour and dynamics of the household is important in understanding factors influencing the criteria people use in evaluating crops. At the household level, the division of tasks and responsibilities was found to be gender-specific and gives rise to different observations concerning selection of varieties. In Lombok and Sumbawa, men are responsible for crop production activities. Women carry out post harvest-activities including drying, cleaning and storage as well as household-level processing and food preparation.

It was realized from the beginning of the study that factors outside the household and production domain of the farmer influence the selection of particular crops. The selection of a variety is embedded in the food production system which may extend beyond the farm household. In commercial farming, aspects of perishability, transportability, and product quality defined by the market (either for fresh consumption or for processing) and reflected in prices influence farmers' choice. Therefore, the interviewing of key persons further along in the food system such as traders, marketers, retailers and agro-processors provides essential information to understand farmers' preferences for varieties. Thus in the course of the research it became apparent that farmers growing the cassava clone *Kuning* did not know the traders' motivation for exclusively buying roots of *Kuning*. Only by interviewing traders and processors were these reasons revealed.

Field observations were an invaluable part of research. Identification and determination of the varieties/landraces in the field proved necessary to obtain a detailed picture of the prevalent varieties, to verify that a variety identified with a certain name is identical for farmer and researcher, and to identify the genetic diversity within a landrace mentioned with a certain name. In the field of post-harvest handling, storage and food preparation, observations provided useful information for understanding farmers' preferences. Illustration materials such as colour photos of pests and diseases and seed samples were also important in eliciting reactions from the farmers which were otherwise difficult to obtain.

A historical component was included in the interview checklist. Since commercialization of agriculture is a relatively recent phenomenon in West Nusa Tenggara, many commercial farmers could also provide information on variety selection in subsistence systems. A relationship between the shift from subsistence farming to commercial farming and changes in variety selection was revealed. The historical information also showed that farmers are receptive to new varieties in all four crops, whether these are traditional varieties from other areas or officially released improved varieties.

Systematically-collected information of farmers' knowledge of crops in genebanks is useful in indicating certain characteristics which are hard to evaluate, such as tolerance to waterlogging, tolerance to drought, suitability for local food products, and many others. As evaluation is a major bottleneck in many genebanks, farmers' information also can function as a first indication of the characteristics present in the germ-plasm collection. The combined collection of germ-plasm and associated farmers' knowledge is particularly useful where funds and manpower for evaluation are limited and genetic diversity in the field is wide.

Due to unforeseen circumstances, the research described has not been able to progress much beyond the descriptive and diagnostic stage. A method should now be developed for systematically classifying and storing farmers' information on genetic material in a databank parallel to passport data. It is intended that the information be verified by on-farm, on-station and laboratory testing of both agronomic and product quality aspects. Promising genetic material has been included in ongoing evaluation and breeding programmes (soybean with some resistance to stinkbugs) and research on storability, starch and sugar contents was started on cassava and sweet potato.

Methods to secure farmers' participation in the evaluation of promising material from formal breeding programmes are being explored. A field day was organized in a village where sweet potato is produced commercially. Farmers were requested to evaluate ten promising varieties of MARIF's sweet potato breeding programme against the dominant local variety on both agronomic and product quality characteristics. For their part, formal breeding programmes are to adapt the criteria they use in varietal selection to reflect more fully the requirements of farmers.

Participatory approaches to management of local resources in South India

James Mascarenhas*

This paper describes how Myrada, an NGO in South India, has developed a highly participatory programme for micro-watershed development, the PIDOW Gulbarga project. It examines the community organization approach taken and the introduction and application of participatory rural appraisal (PRA) methods to local level planning. Some of the ethical issues encountered by the project are discussed. Finally, the paper shows how the approach is now being introduced into some government organizations with a surprising degree of success.

In India, particularly South India, the past few years have witnessed a growing movement towards people's involvement in the restoration and management of natural resources. Much of this activity has been brought about by the pioneering work of NGOs, particularly Myrada. Today there are several participatory projects, each aiming to build up rural people's capacities to be effective partners in the development and management of natural resources. The programmes are mainly concerned with the rehabilitation of wastelands, ancient tank irrigation systems, degraded forests and micro-watersheds. This chapter focuses on experiences from micro-watershed development programmes in the semi-arid and drought-prone areas of the southern states of Karnataka and Andhra Pradesh. It examines the programmes' community organization approaches, or the building up of effective people's institutions within the watersheds, and the introduction and application of participatory rural appraisal (PRA) methods to micro-watershed development.

Organization of watershed groups

From the point of view of programme management, the shift from 'village' to 'watershed' as the unit of development was significant because it sharpened its focus on an integrated and scientific approach to developing the habitats of communities as a way of achieving their own development. But when it came to operationalizing this approach, it was realised that unless the communities themselves were better organized and developed, real and sustainable habitat development could not take place. Since the initial thrust of watershed development does not necessarily correspond to the village land area, this gives

---

* James Mascarenhas dedicates his paper to the farmers of Kamlapur, particularly the farmers of Kalmandargi and Limbu watersheds from whom he learnt so much.

rise to complexities in organizing the community into functional groups within the watershed.

The Participative and Integrated Development of Watersheds (PIDOW) Gulbarga project, a collaboration between Swiss Development Cooperation, the Government of India and Myrada, a large NGO based in South India, gives an idea of the nature of this complexity. Here, several large groups (of up to 136 farmers) had to divide into smaller groups to become functional. This process is common in watershed projects and is facilitated by NGOs who encourage the groups to meet regularly (weekly or fortnightly) in the evenings. These evening dialogues are directed not only towards building up awareness of watershed ecology and resource management, but also towards an understanding of new types of institutions being developed, their roles and their functions. The meetings help foster cohesiveness and co-operation. With more vulnerable groups such as women, tribal and landless people, the non-formal education programme aims to increase awareness about their rights and deprivation and to assist them to realize their own potential and break out of the existing situation.

Credit management programmes

Among the different roles that watershed groups play is the important one of 'credit management'. Credit is a widespread and critical need of people in the area. Watershed groups are encouraged to promote savings by their members. This creates resources in the form of a common fund and also encourages the habit of thrift among the people. Slowly, the situation is changing from one where small and marginal farmers and landless people borrowed from money lenders and landlords at exorbitant interest rates, to one where they are able to get credit through their own corpus funds. Capital formation in the watershed groups is accelerated through matching grants for savings incentives, mobilization of grants and subsidies from the government, and lobbying with banks for group loans. Various income-generating programmes and the creation of individual and group assets in the form of trees, soil and water conservation structures and other land-productivity improvements further encourage this process of capital formation.

People are facilitated to manage credit on their own. Their management systems consist of a combination of their own rules and regulations and traditional systems supplemented by some 'outside' systems such as cash books and registers, and book keeping and accounting tools, for which some training is given. Experiences show that the common fund tends to be managed very efficiently by its members in terms of prioritizing borrowers, determining legitimate purposes for loans, amounts to lend and interest to be charged, and in recovering loans.

Credit management is a core activity of the watershed groups and is extremely important in both restoring the resources of the watershed and in managing them in a creative and sustainable way. The groups also become instruments of change. With better organization, awareness and increased confidence, they begin to place demands on the government system, and form an appropriate mechanism by which government inputs can be channelled and managed for development of rural areas.

Participatory rural appraisal methods in natural resources management

Since the first trial on the use of rapid rural appraisal methods in 1989 under the Myrada PIDOW Gulbarga Project, these and participatory rural appraisal (PRA) methods have provided the framework and tools for analysis and understanding of rural people and their environment. A typical PRA exercise for planning a watershed development programme is held over three days:

Day I: Warm up

The objective of the first day is to establish a rapport between the project and the village. Outsiders familiarize themselves with the village, the villagers and the work which villagers do. Sometimes they also engage in common village tasks like helping in the harvest, learning to build houses and so on. The stress is on attitudes and on appropriate interviewing. Villagers are briefed about the programme and consulted on timing of the discussions and so on. The outsiders identify key resource persons among the villagers. A few preliminary exercises are held which may include *time lines, trend diagrams* and *historical transects.* Sometimes *seasonality* or *matrix-ranking exercises* are initiated. The seasonality exercise is important, because it indicates periods when villagers may be free to carry out possible additional activities, for example forestry. This kind of 'fine tuning' is essential from the point of view of people's participation in the programme. Similarly, the matrix-ranking exercises for trees helps to establish the species mix in any forestry programme according to the villagers' preferences.

Day II: Exploratory

On the second day, groups of outsiders and villagers enters into a detailed study of the watershed. *Sweeping* or *combing transects* are carried out, whereby groups of outsiders comb different sections of the watershed with farmers who have land in that area. In the case of forests, which are invariably populated by tribal people, the study is done with tribal villagers as guides.

Transects. help to locate and facilitate the discussion of problems and opportunities with each individual farmer on his or her land and in the watershed as a whole. In many cases it has been found that land records are out of date or that farmers do not have the title deeds for the land they are cultivating. This assumes particular significance in government-implemented programmes where details of survey members, farmers names etc are required. Another frequent problem is that of management of the 'commons'. Villagers will not participate in the development and management of these lands unless usufruct rights are well defined and titles clear. The transect group's combined knowledge of the technologies, particularly those relating to conservation and management of resources, enables it to arrive at a 'treatment plan' covering soil and water conservation, forestry, agriculture, horticulture etc. for the watershed. These treatment plans are indicated on a 'map'.

Mapping is the high point of the PRA exercise. Usually the map is prepared on the ground and later copied onto paper for record purposes. In the mapping exercise, each group represents on the map the treatment plan arrived at through the transect. Such visual representation stimulates discussion on issues such as what is currently happening in terms of managing the resources, what should happen, problems, opportunities, constraints and so on. A strategy for treating and managing the watershed is also discussed and arrived at. Mapping exercises have been extremely powerful for both outsiders and villagers in understanding the dynamics of watershed management and visualizing its development.

Day III: Concluding
On the third day the treatment plan is finalized. The 'final' plan is a consensus between the villagers and the outsiders on what is required and how it is to be done. Generally, the plan has five components:

(1) Treatment plan (e.g. soil and water conservation works, forestry);
(2) Budget plan (including community contributions in cash, kind or labour);
(3) Time plan (scheduling work according to the villagers' calenders)
(4) Implementation plan (roles and responsibilities of various agencies including the government);
(5) Management plan (roles and responsibilities of each party).

Building on PRA

Participatory rural appraisal methods have provided outsiders with a clearer understanding of rural people's ways of looking at things. The staff have begun to see more and discovered many things they had not previously seen even though they may have been working in the area for several years. Significant among these discoveries were indigenous technologies and traditional systems relating to management of natural resources. We discovered how the environment has impacted on people's lives and how they have adapted to it.

For example, the seasonality exercise told a whole story enabling us to see patterns in cropping, agricultural operations, labour employment migration, income and expenditure, debt and credit, fodder and milk, human and cattle diseases, and so on, all in relation to each other and to the agroclimatic conditions. During one PRA exercise we uncovered a seasonality of oestrus or heat in cattle. This ensured calving in October/November when there was plenty of fodder available and milk yields could be maximized. This was critical information for the cross-breeding cattle programme.

While doing a time-line exercise in another village we came to know that the 1972 drought was an important event in its history. There were several major consequences of the drought as given by the farmers:

o 'Conditions were so bad that we had to go far away from our village in search of work or even a little food. Some of us reached as far as Bombay. Because of this, we have lost our fear of the 'unknown'. Nowadays it is common practice for all able-bodied persons to migrate to Bombay after the agricultural season, where we earn good money as construction labourers.'

o 'We cut all our trees and sold them because we had to feed ourselves and our families.'

o 'Our local sorghum variety was wiped out during the earhead stage. Because we couldn't get seed the next year, we were forced to switch to hybrids. Hybrids do not give us as much straw as our local varieties did. Therefore we were forced to extend our cultivated areas to include grazing lands. Because the grazing land was reduced we were forced to graze our cattle in the forests.'

o 'Because our net cultivated area has increased, we cannot apply as much farmyard manure to each field as we used to. Since these hybrid varieties require more fertilizer we are applying larger quantities of chemical fertilizer.'

Historical transects give trends in resource use and management. These and other methods help to locate discussion points from where interventions and actions can be discussed and negotiated with the local communities, leading to courses of action that are more relevant, better more easily managed by the people and more sustainable.

Many more illustrations could be given of how the knowledge of the community about its environment finds expression through PRA. In most cases, local knowledge is collective and accumulated over several generations forming an amazing knowledge pool. This includes an incredible number of indigenous technologies and traditional management systems, most of which are lying undiscovered and not yet understood.

The knowledge pool and the creativity that exists within rural people is an untapped resource in rural development and natural resource management. This is of particular significance because such indigenous technologies and management systems are low cost, appropriate to the situation and easily managed by the people. This does not mean that no outside technology or interventions are needed. They may be: but they have to be appropriate and carried out within a suitable framework. This is illustrated by Figure 3.3: 'Four squares of knowledge'.

Unfortunately, we the 'outsiders' have been stuck in mode 2 for too long. This has affected the development of rural people: retarding them and destroying their capabilities and confidence in themselves and in their knowledge and systems. There is an urgent need to change this to the mode indicated in square 3, where we begin to discover and appreciate what people know. Validation of local knowledge is required but so is validation of 'our' technologies in 'their' context and situation. Following this approach of fostering innovations on either side, of finding what works and what is appropriate, will result in an enlarged square 1, which can also be called the 'square of common and shared understanding' or the square of *participatory technology development* (Figure 3.4). This square represents the 'basket of choices' from which village people can choose technologies or programmes according to their capacity and ability to manage them.

Example:			Example:
- Water is good for crops - Fertilizer increases yields - Soil gets eroded by rain (especially heavy rains) etc.	**1** We know They know	**2** We know They don't know	- Advanced meteorology forecasts etc - Nutrient composition of fertilizers micro-elements etc.
Example: - Productivity of different plots of lands - Local fodders - Local problems - Traditional farming systems - Medicinal plants and local medicine etc.	**3** We don't know They know	**4** We don't know They don't know	Example: The future: - What new varieties will be developed - How will these perform _ Weather conditions etc.

Figure 3.3: *Four squares of knowledge*

Challenges for PRA

Most of the learning about PRA methods and their application, and about rural communities, their knowledge and interaction with their environment, took place where an NGO was already working. The NGO presence and the existing rapport provided an environment which enhanced the villagers' participation, information generation and consequently the outsiders' learning. The lessons learnt in these settings have enabled us to develop applications in various areas from health and nutrition to water and sanitation, credit, animal husbandry and fisheries.

In several instances, projects were changed mid-course as a result of what the people explained to us. For example, one project was set on carrying out a programme for de-silting tanks to raise their water storage capacity, so as to increase the area under irrigation in the tank command. We found out that the majority of the villagers had land in the catchment and they insisted that treatment of the catchment, particularly the upper slopes, was more important than de-salting the tank. In any case, they said, a process of de-silting was already being carried out by farmers, who put silt on their fields to increase the fertility, and by brickmakers and contractors who used the accumulated sand and clay for building construction.

Ethics

Questions have arisen over the ethics of conducting PRA exercises. Are we not raising people's expectations? Are we not taking up their time? Were they not participating out of a sense of duty, because they felt obliged to the outside agency?

133

Our technologies and knowledge + Their technologies and knowledge Square of common knowledge and understanding Basket of choices and technologies	

Figure 3.4: *The square of common and shared understanding*

Attempts to resolve these questions are still being made. However, certain clear distinctions can be made. For example, in cases where the PRA was purely for outsiders to learn, we resolved to compensate the villagers for their time, by making a donation to the temple, mosque, school or village common fund. In cases where the purpose of the exercise was for planning the development of the village or watershed, it was generally agreed that the villagers needed to give time and participate. However, in both cases serious efforts are being made to set the timing of village visits at the convenience of the villagers and to share food and snacks with them. On several occasions the villagers shared their food with the outsiders. The outside participants were asked to be sensitive and alert to possible needs of the villagers and opportunities for development, without raising expectations. In every case these exercises resulted in a great deal of learning for the outsiders.

Other questions arising are 'authorship' of information - who should get the credit if new technologies or systems are discovered? Another issue concerns 'jargonization' or 'mystification' of the methods and the interpretation of data. Openness and demystification of methods are essential if villagers are to benefit fully.

Finally, we need to address the issues of 'quality control' and the 'legitimacy' and 'spread' of PRA methods, and how to ensure that these are achieved. An important realization by our field staff has been that 'rapid' cannot be 'participatory'. We much preferred the participatory mode of rural appraisal in which there was greater scope for interaction between the villagers and for them to participate in the development process. But we realise that in terms of scaling-up the approach, the method has to be both rapid and participatory.

From planning to implementation

PRA methodology has been introduced into several government organizations in South India such as the Dry Land Development Board of Karnataka State and the Drought Prone Area Programme of Andhra Pradesh, for the micro-planning of watersheds. In these exercises the learning has been of a different nature, giving an idea of how the method works in the government system. A major achievement has been that watershed maps are no longer prepared on topo sheets or cadastral maps in the offices at headquarters, but out in the villages with the people. These exercises still have to be followed up with community organization during the implementation phase to sustain participation in the programme.

In several places a stage has been reached where the implementation of watershed development activities has been handed over to the people, with outsiders lending support rather than causing interference. This has involved a major shift away from planning and implementation by outsiders. In this way, villagers' confidence has been developed and the local economy has benefited with cash flowing to the people in the watershed, instead of to outside contractors and other vested interests. The watershed groups are crucial in this process, taking responsibility for managing the implementation of the programme and the assets created. Recently a few project evaluations have been carried out using this approach, and the idea is spreading.

Towards long-term sustainability

We have seen how, in South India, this 'new' approach of enabling rural people to participate in their own development enhances the learning of both outsiders and villagers. Such learning about indigenous technologies and traditional ways in which people are managing their resources helps outside agencies to define their role, and to understand the role that rural people have to play in development. It shows the need to make use of rural people and their knowledge as a resource in the development of rural areas. If this approach is accepted it will no doubt have an enormous impact on enhanced participation from the villagers.

PRA is not a 'one shot' affair. It is the start of a process a process of learning from and with rural people about their environment, their technologies and their systems of management. Hopefully, it is a process towards more sustainable development, provided it is strengthened by a consistent and responsive engagement of outside agencies with the rural people. This calls for empowering rural people through better organization, and strengthening their capabilities through the development of appropriate and effective people's institutions with authority and responsibility for managing their resources and environment. To achieve this, support will be needed from all outside agencies: non-governmental, donor and, especially, government.

Linking local knowledge systems and agricultural research: the role of NGOs

Kate Wellard*

This chapter considers the different ways in which NGOs link local people and formal sector research and extension organizations to access and promote local knowledge. It is based on the findings of a three-year study of NGOs and agricultural technologies carried out by the Overseas Development Institute (ODI) with examples taken mainly from Africa. Factors influencing the development of local knowledge and its incorporation into 'mainstream' agricultural research are then considered, including the ideological leaning of the NGO, the motivations of researchers, the wider macro-political economy and the roles played by donors. Finally, ways are suggested in which NGOs might further strengthen local knowledge systems.

Over the last two decades, non-governmental organizations (NGOs) or non-profitmaking development bodies have emerged as prominent institutions alongside national governments and official donor organizations in many developing countries. Three main types of NGO can be discerned: those acting mainly as a channel of funds to development projects; professional NGOs which provide services such as training or research for other organizations; and grassroots service organizations (GRSOs) which are local organizations with an in-house capacity to provide some services to their members. NGOs are highly heterogeneous with important differences in, for example: their philosophy and view of the development process; their source of funding; whether they are North- or South-based; the degree of control accorded to client farmers and groups over agenda-setting; and their interest and level of expertise in developing agricultural technologies.

NGOs have been involved in most areas of agricultural and environmental technology development. Where government services have been weak, as in much of Africa, NGOs have often provided inputs and supported extension for fairly conventional programmes. Increasingly, however, NGOs have been carving out their own roles which may differ from, or conflict with, government programmes. In Africa, current areas of NGO interest include:

○ indigenous tree species, management practices and uses (including agroforestry);

○ local varieties of food grains, particularly in the face of genetic erosion due to drought and promotional campaigns for modern varieties;

---

* Kate Wellard dedicates her chapter to the farmers of Makunduchi in Zanzibar.

o local varieties and uses of under-utilized crops, including vegetables;

o soil and water conservation practices;

o ethno-veterinary medicine and management practices of livestock keepers and herders;

o work with disadvantaged groups and communities, including women, tribal groups, and disabled farmers.

This chapter considers the different ways in which NGOs link between local people and formal sector research and extension organizations to access and promote local knowledge. It is based on the findings of a three-year study of NGOs and agricultural technologies carried out by the Overseas Development Institute (ODI) with examples taken mainly from Africa. Factors influencing the development of local knowledge and its incorporation into 'mainstream' agricultural research are then considered, including the motivations of researchers, the wider macro-political economy and the roles played by donors. Finally, ways are suggested in which NGOs might further strengthen local knowledge systems (LKS).

How NGOs link with local knowledge systems

The involvement of NGOs in local knowledge systems is varied and complex. Some work to uncover knowledge and incorporate it directly into a development project: local knowledge is often an important component of NGO action-research projects. Others try to link farmers and farmer groups to researchers and extensionists in a bid to have their needs addressed and to encourage the development of technologies based on local resources and knowledge.

Figure 3.5 shows the main roles played by NGOs in linking the different actors in local knowledge systems. In contrast to the orthodox technology generation and transfer model where knowledge is presumed to flow from researcher to extensionist to farmer, farmers, whether individually or through their groups, are seen to play a critical role in contributing local knowledge to the research process.

NGOs, farmers and their organizations

There are at least four ways in which NGOs link with farmers and farmer groups to develop local knowledge systems. First, NGOs accumulate local knowledge by close observation of farmers and their practices. This is facilitated by NGO staff living and working in rural communities. Other location-specific knowledge may be sought through more formal methods such as participatory rural appraisal (PRA). NGOs have played an important part in developing approaches which confer respect and control to the custodians of this knowledge and some have even persuaded government to use such approaches (see Mascarenhas, Part Three).

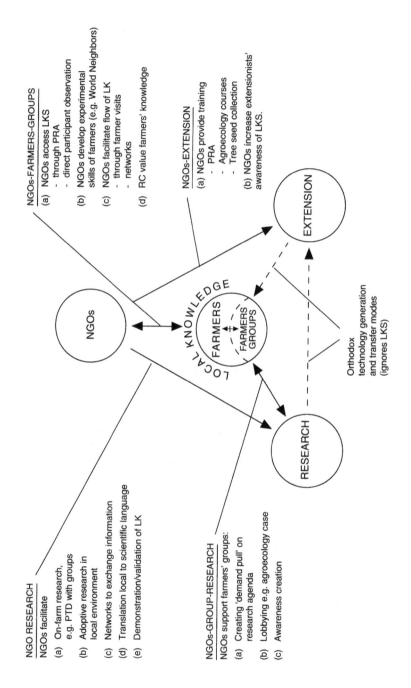

NGOs-FARMERS-GROUPS

(a) NGOs access LKS
   - through PRA
   - direct participant observation

(b) NGOs develop experimental skills of farmers (e.g. World Neighbors)

(c) NGOs facilitate flow of LK
   - through farmer visits
   - networks

(d) RC value farmers' knowledge

NGOs-EXTENSION

(a) NGOs provide training
   - PRA
   - Agroecology courses
   - Tree seed collection

(b) NGOs increase extensionists' awareness of LKS.

EXTENSION

LOCAL KNOWLEDGE

FARMERS

FARMERS GROUPS

NGOs

RESEARCH

Orthodox technology generation and transfer modes (ignores LKS)

NGO RESEARCH

NGOs facilitate

(a) On-farm research, e.g. PTD with groups

(b) Adoptive research in local environment

(c) Networks to exchange information

(d) Translation local to scientific language

(e) Demonstration/validation of LK

NGOs-GROUP-RESEARCH

NGOs support farmers' groups:

(a) Creating 'demand pull' on research agenda

(b) Lobbying e.g. agroecology case

(c) Awareness creation

Figure 3.5: *NGO links with local knowledge systems*

138

Second, NGOs help develop farmers' analytical, experimental or communication skills. World Neighbors is one NGO which has recognized that agricultural research and extension are unable to meet the needs of the majority of small farmers. It has therefore adopted an 'agricultural self-development' approach: strengthening farmers' existing capacity to analyse agricultural problems and make appropriate changes to their farming system themselves (Gubbels, 1988). The Information Centre for Low External Input and Sustainable Agriculture (ILEIA) has documented the experiences of dozens of programmes supporting farmers' research and has synthesized these into a series of steps which they term 'participatory technology development' (PTD). Many of the experiences are those of NGOs.

Third, NGOs have used their links with farmer groups in different locations to facilitate a flow of information and knowledge between farmers. Farmer exchange visits and networks have proved highly successful means of introducing farmers to new technologies and exchanging ideas. In the High Andes, traditional seed fairs are being supported by both researchers and NGOs as a means of strengthening local diversity and capacity to experiment (Tapia and Rosas, Section Three).

Finally, GRSOs such as the Organization of Rural Associations for Progress (ORAP) in Zimbabwe are linking with local community organizations, encouraging members to question the development process of which they are part and to take control of their own development (Ndiweni, 1993). Local people are beginning to examine critically the roles and assumptions of outsiders and to reassert their own beliefs and values. The championing or revaluation of local people's knowledge is central to this.

NGOs and extension

Whilst many NGOs have undoubtedly used traditional extension approaches, characterized by the transfer of modern 'scientific' methods and inputs to farmers, others have taken the lead in fostering a more participatory approach to developing and promoting agricultural technologies based on local knowledge. Surveys of NGOs in two districts in Kenya (Charles and Wellard, 1993; Kaluli, 1993) revealed that the majority were promoting some programmes based around locally-discovered technologies (frequently indigenous trees or vegetables), while two were genuinely involving local people and their knowledge.

By becoming involved in training and orientation courses for government and NGO extensionists, NGOs have been able to impart experiences and raise awareness of local knowledge at a wider institutional level. Kenya Energy and Environment Organizations (KENGO) trains extensionists in the collection, preservation and uses of indigenous tree seeds, whilst the Latin American Consortium for Agroecology and Development (CLADES) collects and synthesizes information on agroecological activities for training NGO and government technicians (Bebbington and Thiele, 1993).

NGOs and research

NGOs also assist in raising researchers' awareness of local knowledge, demonstrating its validity. Much of their work has been in non-traditional food crops and sustainable agriculture where there has been little official research interest until recently. NGO research has been mainly applied and adaptive, reflecting concerns for rapid and tangible results.

NGOs can facilitate on-farm, participatory research by organizing farmers and liaising with researchers, where necessary providing support such as transport to the field. Langbensi Agricultural Station in Northern Ghana is an NGO which is stimulating collaboration between farmers and the local government research station using the PTD approach (Wellard and Kolbilla, 1993). As a result, local knowledge on soil fertility technologies and on host plants for controlling the weed *striga* has been incorporated into field trials and farmer assessments are being included in the evaluation process. In North Omo, Ethiopia, where there are a large number of development organizations and a growing recognition of the need to develop low-input technologies, FARM-Africa has begun training NGOs and government research institutes to carry out farmer participatory research.

A number of NGOs have taken on the task of documentating and disseminating of information on local knowledge systems. International and South-based NGOs have established information exchange networks on seeds and plant breeding, traditional crops, indigenous trees and agroforestry, and organic or low-external input agriculture. Such information may be used by researchers and practitioners as an input into field programmes, or for *in situ* genetic resource collections which are maintained by the NGO and farmers themselves. NGOs have also played a role in bridging the gap between formal and informal knowledge systems by documenting and promoting the use of local classifications and uses of soils, farming systems and plants, and sometimes in 'translating' these into scientific terminology.

NGOs link farmer groups with research

Some NGOs feel that simply demonstrating the existence and use of local knowledge systems is not enough to secure its wide acceptance as legitimate knowledge amongst the formal sector, nor to ensure that the custodians of local knowledge receive adequate recognition. The best way to achieve this, they feel, is to strengthen grassroots organizations to lobby for services and support from agricultural research institutions and for their interests to be represented. Support for local knowledge and values has thus become closely identified with movements to increase representation and democratization of disadvantaged rural people in many parts of Africa and Latin America.

Influences on the development and spread of local knowledge

Interest in local knowledge appears to be on the increase with popular concern about the environmental impact of chemical inputs in agriculture and the destruction of forests which are the source of many useful plants. Observers within and outside the NGO community have noted the failure of the Green Revolution, the replacement of traditional practices with new scientific techniques and inputs, to generate sustained increases in agricultural production and improvements in livelihoods, particularly of resource-poor farmers in difficult environments. They stress the dangers of reliance on purchased inputs (that are frequently not available locally) and on practices which can cause environmental degradation, and the tendency of such approaches to increase the dependency of rural people on outsiders who control information and make investment decisions.

These concerns have led to a search for alternative approaches to development emphasizing the environment as the basis for sustainable livelihoods, with local technical knowledge as a central component (Altieri and Yurjevic, 1989).

Official and NGO donors are making significant contributions to this process. The equation of local knowledge with sustainable development and environmental protection has produced large inflows of aid. In many cases existing NGO efforts have been supported, whilst other projects, including several by government researchers, appear to have been stimulated by the availability of funds. It is too early to tell, however, whether these developments are having the intended impact on the welfare and status of rural people.

Deteriorating economic conditions in many developing countries, particularly Africa and Latin America, and the cut-backs in government expenditure under structural adjustment programmes appear to be having an effect on efforts to develop local knowledge. First, increases in prices of imported inputs have fuelled the search for technologies based on locally-available materials and associated techniques. Second, researchers, faced with drastically reduced budgets, are beginning to look for new partners: frequently these are NGOs with a strong field presence and close links with farmers. Interest in local knowledge has come from universities and some of the newer, more flexible government departments experiencing a dearth of ready-made technologies-departments such as forestry and agroforestry (see Buck, 1993).

Against the factors contributing favourably to the development of local knowledge systems must be set the continuing threat of their exploitation by commercial interests. Retaining control of knowledge within communities is seen as critically important in maintaining the productivity of knowledge and rewarding its holders who are often women or dwellers of marginal areas and amongst the poorest people in society. NGOs such as the African Centre for Technology Studies (ACTS), RAFI and GRAIN are active in lobbying governments and international organizations on property rights in genetic materials, whilst consumer organizations are working to increase awareness about local products and their producers.

Linking local knowledge systems and agricultural research: future roles for NGOs

NGOs have already had some success in linking farmers and their organizations with researchers and extensionists to strengthen local knowledge systems. In future these might be developed further by:

o assisting researchers to identify and collaborate with particularly knowledgable farmers; as well as supporting the needs of under-represented groups to researchers (including women);

o documenting and disseminating findings and methodologies on local knowledge (and demonstrating their viability) in scientific journals; and o facilitating exchange of information and ideas between the groups, in farmers' fields as well as through workshops.

Perhaps the most important finding of the ODI study on linkages between farmers, NGOs and researchers in developing agricultural technologies is that whilst there might be potential advantages to be gained from interaction, it will not be productive unless all sides perceive benefits and an atmosphere of trust prevails. Mutual respect is essential for researcher-farmer collaboration in developing technologies based on local knowledge. NGOs, either as researchers, catalysts or lobbyists, can have an important and sensitive role in bringing local knowledge into the forefront of agricultural research and policy.

Conservation and utilization of food and medicinal plants in Botswana

Martin N. Mbewe*

This short paper describes the cultivation and uses of veld products based on knowledge held by the rural and remote area dwellers (bushmen) of the Kalahari Desert and documented after extensive dialogues with anthropologists. Many of the products have been over-exploited and some are nearing extinction. The Botswanan NGO, Thusano Lefatsheng, is working to reverse this trend by encouraging the rural and remote area dwellers to cultivate plants which have hitherto been considered wild, and by promoting the economically and socially sustainable utilization of veld products.

The rural and remote area dwellers, Basarwa (bushmen), of the Kalahari Desert use a variety of natural resources for food, medicines, crafts, thatching and, since the introduction of money, as sources of income. Local knowledge of these veld products has been documented followng extensive dialogues between anthropologists and local people and forms the basis of work on utilization and conservation by a Botswanan NGO, Thusano Lefatsheng ('Helping each other in the country').

Local knowledge of food plants

The morama bean *(Tylosema esculentum, Fabaceae)* is a herbaceous runner native to the arid and semi-arid grasslands of Southern Africa. In Botswana, it is abundant in the sandveld, especially in the Ghanzi district of the western part of the country. Traditionally, the dry beans are roasted and eaten with sugar or salt. Young green morama beans are eaten raw and sometimes boiled and ground to make a porridge or drink. The young tubers are traditionally eaten fresh as emergency sources of water for humans and animals during droughts. They can also be baked, boiled or roasted and eaten as such. The stems, leaves and seeds, when left on the plant, provide food for cattle, goats and wild animals (Mbewe, 1992a).

In an attempt to understand why local people eat the beans, the Food Technology Research Service of the Botswana Technology Centre (BTC 1991) have discovered that they are rich in protein (13.9 per cent: whole nut), oil (14.9 per cent), carbohydrates and minerals. Protein content is roughly similar to that of soybeans, but quality has been found to be slightly better and is comparable

---

* Martin Mbewe dedicates his paper to the Basarwa (Kalahari Bushmen), on whose knowledge the paper is based.

to casein or milk proteins. Kernel oil content is about twice that of soybeans and approaches that of groundnuts (Mbewe, 1992b). Further studies have established that morama nuts could be processed into snacks and butter on a commercial scale and that large-scale production of oil is technically possible (BTC, 1991).

The morula tree (*Sclerocarya caffra*) is a member of the Anacardiaceae family and is found in the south-east, eastern and northern regions of Botswana. It is widely distributed throughout Africa from South Africa to Ethiopia and Sudan. The morula fruit is traditionally eaten fresh but it can also be dried and stored for later use. It is sometimes cooked in porridge to add flavour or cooked with vegetables or meat and may be pounded and formed into cakes (Shone, 1979). It is also used in the preparation of high-potent beer; as an insecticide and a germicide.

The mongongo tree (*Ricinodendron rautenenni*) is a fairly large deciduous tree which is indigenous to Botswana. It is found in the northern parts of the country and in northern Namibia, southern Angola and Zambia, western Zimbabwe, northern Mozambique and the southern tip of Malawi. Traditionally, the sweet flesh of the mongongo fruit is eaten fresh and scientific nutritional analysis have shown that it contains about thirty per cent sucrose, nine per cent protein, minerals (especially potassium), B vitamins and some vitamin C. The spongy flesh may be mixed with raw cereal to prepare porridge. The nut's kernel has been found to contain fifty seven per cent oil which could be used in the manufacture of both paint and cooking oil (Wehmeyer, 1980).

Local knowledge of medicinal plants

In Botswana, grapple (*Harpagophytum procumbens* DC) occurs mostly on disturbed sites of the Kalahari sands. It is also found in parts of Namibia and South Africa. In traditional medicine the tubers of the grapple plant are sliced, dried and crushed into granules. Boiling water is added and left to infuse overnight. The following morning the liquid is decanted and the resulting portion is drunk in three equal portions through the day. The medicine is considered to 'clean the blood'. It is used in the cure of hypertension, high blood pressure, diabetes and stomach disorders. Active ingredients (harpagoside, harpagide and procumbide) have been isolated by Western medical researchers and have been proved effective against arthritis and rheumatism (Moss and Taylor, 1981). The tablets or capsules made from these plant extracts are also recommended as medicines for general health.

Lengana (*Artemisia afra*) is a woody perennial herb which occurs in the highland areas of eastern and southern Africa at altitudes of 1500-2000 metres above sea level. The plant grows along most river valleys in Botswana. In traditional medicine lengana leaves are used as a herbal tea believed to be effective against flu, measles, coughs, chills, malaria, loss of appetite, stomach ache and other gastric disorders. The leaves are also mixed with oil as perfume. They have been found to contain an essential oil whose major constituents are cineole, thujone, isothujone and camphor. Lengana roots are boiled and the decoction drunk as a cure for internal worms.

Musukudu (*Lippia javanica*) is a deciduous perennial plant of the verbenacea family which grows well on Botswana's hardveld. Traditional medicine uses musukudu leaves as a herbal tea believed to be effective against some stomach disorders. The tea has a good taste and a pleasant aroma.

Musukujane (*Lippia scaberrima*) is another deciduous perennial plant of the verbenaceae family. It is indigenous to Botswana and occurs mostly along roadsides and in woody grasslands in the hardveld. Musukujane leaves are used as a herbal tea for coughs, colds, measles, malaria, dysenteric fever, nausea, indigestion, flatulence, palpitations and vertigo. The leaves contain a crystalline alcohol called lippianol which could have a commercial use.

Building on local knowledge of veld products

All these veld products are being exploited in their wild state. Some of them such as grapple and morama have been so over-exploited that they are now nearing extinction. Harvesters have to walk long distances in search of products which not long ago were close to their settlements. In an effort to conserve what remains in the wild, the government of Botswana has declared grapple a legally-protected plant which can only be harvested and sold with a permit. Although this has somewhat reduced the rate at which grapple is being exploited, it has not helped in increasing the plant population. In fact, it has continued to dwindle.

While Thusano Lefatsheng views legal protection as a significant step towards preventing plant extinction, it believes that cultivation of these plants is the ultimate solution. To this end, agricultural research at Thusego, Thusano Lefatsheng's research station, is geared towards the domestication and eventual development of production technologies for veld products. Ecologically-sustainable harvesting methods which will ensure the continued regeneration of perennial plants are also being developed. The organization's Extension Department is working with poor rural and remote area dwellers encouraging them to cultivate plants which have hitherto been considered wild. Working with Thusano Lefatsheng's Commercial Division (Processing and Marketing), it must also ensure that utilization is economically and socially sustainable.

In addition to the agronomic research underway, *ex situ* methods of conservation are in use. Extensive collections of what remains of the over-exploited veld products are planned. So far limited collections have been made and seeds are in storage. Field plantings currently being established will not only ensure the continuous collection of scientific data but will also be a source of seeds for the future.

Interventions and sustainable agriculture in South Africa

Dan Taylor

This paper describes the attempts of an NGO, CLIARD, to develop appropriate research models to promote sustainable agriculture among oppressed and marginalized black farmers in South Africa, while recognising the fact that dominant agricultural research models are rooted in a political and socio-economic system which is against the interests of black farmers and seeks to undermine them. This involves revalidating the traditional practices of black farmers, building upon their innovative and adaptive traditions, and facilitating local organization through which farmers can engage with state structures and make demands for their needs to be met.

Black agriculture in South Africa has been characterized by a history of land dispossession, a legacy whereby the agricultural production of black farmers has been marginalized, and by a long process which has involved inculcating into the minds of farmers the idea that traditional practices are inherently inferior. Land dispossession and marginalization have been overt, with direct and obvious steps having been taken by a white minority before and after the declaration of the Union of South Africa in 1910. The rejection of traditional practices have been covert, with the process given added momentum by the black agricultural personnel indoctrinated into monocultural practice.

Black South Africans, despite the repeal of much discriminatory legislation, are for all intents and purposes still confined to 13 per cent of the land in South Africa. The subsidization of white agriculture by a minority government in need of the rural vote, together with access to easy credit and infrastructure support, has given it considerable advantage over black agriculture. The deliberate efforts by the authorities to destroy black agriculture has been documented by Bundy (1988). But it is the more subtle and covert effect whereby black culture, tradition and indigenous knowledge systems have been undermined that may prove to be the most lasting impact of colonization in its many forms, especially the aberration of apartheid.

This chapter begins by describing the type of agriculture promulgated by the South African authorities at all levels, both black and white. Secondly, it offers a definition of sustainable agriculture and links indigenous knowledge systems to the concept of sustainability. A few examples of traditional agricultural practice are described to illustrate its value and to demonstrate that traditional practice is a dynamic system which is innovative and adaptive. Finally, the chapter considers challenges facing researchers; one NGO's response; and institutional support needed for sustainability.

146

The imperialism of white agriculture

The difference to even the casual observer of adjacent black and white landholdings in the rural areas of South Africa is obvious. The former can be identified by small areas under cultivation, lack of infrastructure, highly eroded areas and overgrazing. The latter have large farms, carefully-planned and fenced, contoured lands, some degree of mechanization and carry fewer livestock. Whilst this description may be a generalization and a simplification of the diversity of vegetation types and land-use systems in South Africa, it highlights the point that in the minds of most people, the difference is attributable to the superiority of one type of agricultural system over the other, namely modern systems over traditional practice.

This is false not just in its conclusion, but in the process of deduction by which it is reached: that agricultural problems are only agricultural in nature. This argument views agriculture as a purely biophysical process and would go no further than analysing such factors as climate, soil, slope and cultivar. However, agricultural systems are part of larger systems in which socio-economic and political factors have as much influence on agricultural systems as biophysical ones. Agriculture is a human artefact whose determinants extend beyond the farm perimeter (Altieri, 1987).

The idea that agricultural best-practice is that of modern systems commences at a very young age by the very nature of the subject matter taught at schools as part of the agricultural curricula. The model is that of tractor and harvester, dairy cow and common vegetable. In the agricultural colleges where generations of extension officers have been trained for the black communal areas this process is continued. Best practice becomes the modern agricultural package.

The agricultural extension officer when leaving agricultural college, usually as a fairly inexperienced agriculturist, has only one advantage over the resilient and experienced farmer: his text-book knowledge with its agricultural package of new cultivars, pesticides and inorganic fertilizers. It is with missionary zeal that the new word is preached; or alternatively, faced by the immensity of the task and the bureaucracy of government or parastatal institution he/she represents, nothing is done at all.

The alternative

African ecological systems are noted for their diversity, and the peoples of Africa have practised both extensive and intensive agricultural systems to cope with this ecosystem diversity.

The risk-averse practices of the low-resource farmer underpinned by the need for survival strategies, have necessitated the use of polycultures to increase the probability of a yield during every growing season. Whilst diversity is often achieved at the expense of productivity (at least in the years of above average rainfall), it is in times of relative drought that such a strategy comes into its own with at least some yield being obtained from the diverse crops grown. Multiple-cropping systems are better able to cope with fluctuations or weather variability during the growing season, as a variety of species allows different agroecological niches to be occupied.

However, this analysis should not be taken too far. The practice of monoculture has its own logic. By simplifying agricultural systems as opposed to the diversity of natural ecosystems it is possible for the farmer to attempt to create the optimum environment for the growth of the single crop (only intra-species competition occurs). It is exceedingly difficult to manage a number of crops growing in association owing to inter-species competition. Thus if a high initial outlay such as an irrigation system is required, there may be no alternative to monocropping to realise a return of sufficient magnitude in order to produce the capital and interest repayments.

Clearly, the rationality of monoculture adoption will be a function of the relationship between the price of inputs and the price of outputs and, as such, of state policy. The subsidization of farming inputs by government will favour the adoption of such technologies with the non-adopter incurring an opportunity cost. Furthermore, if the social cost of fertilizer and pesticide use in terms of pollution is to be borne by the farmer rather than society in general, then the equation will change.

The point is that there is no simple solution. However, if agriculture is to be sustainable then the question is, what then are the necessary preconditions for its attainment? It is precisely because we do not know the answer to this question that we need to keep all our options open. It then becomes essential to preserve biodiversity by saving landraces, not because what is old is necessarily best; it is because we cast aside at our peril centuries of plant breeding by traditional farmers. The very survival of landraces is a testament to their success.

Sustainable agriculture needs to be productive, stable, environmentally sound and maintained with a minimum of external inputs, yet at the same time must offer all farmers an equitable return. What this means is that a new orientation is needed. This should build upon the risk-reducing and resource-conserving complex systems of sustainable agriculture derived from indigenous knowledge and practice, but selectively include the advances of modern science and technology (Taylor, 1991a).

Solutions cannot be applied universally because each potential agroecosystem needs its own solution and whilst certain generalizations may be made, these should be accepted with caution. Chambers (1989) has shown that low resource farmers require a 'basket of choices' rather than a 'package of practices'. As risk-averse managers of complex systems, low-resource farmers are incremental adopters of new technologies: they introduce small changes to existing practices and hardly, if ever at all, will a totally new package of practices be introduced immediately (Taylor, 1991b).

Indigenous knowledge systems as dynamic traditional practice

North-eastern Natal is an area with an annual rainfall ranging from 1000mm on the coast to 400mm in some interior lowland savannas. A variety of agricultural systems are practised by black farmers in the KwaZulu homeland (Bantustan) including commercial sugarcane, cotton and timber production. Numerous practices derived from traditional systems of management are still in use but are ignored by agricultural researchers. None of these are static systems but all include a certain degree of new innovations. An understanding of these should

148

allow for the development of more appropriate agricultural technologies for the region (Taylor, 1988).

In the Hlabisa district, in an area receiving an average annual rainfall of 700mm, a farmer has successfully combined traditional and modern practices. This farmer plants a diverse range of crops including maize, sorghum (*Sorghum bicolor* subsp. *cafforum*), bambarra groundnut (*Voandzeia subterranea*), cowpeas (*Vigna unguiculata*), sweet potatoes and modern bean varieties. He continues to plant traditional landrace varieties of maize but in addition uses a hybrid. When asked why he continues to use a traditional variety of significantly lower production potential he replied, 'Why discard something that has served us so well for so many years?'

This farmer is also a ploughing contractor and often accepts kraal (cattle) manure for payment in lieu of services rendered. This he spreads over his fields and complements it with inorganic fertilizer.

In Eastern Maputaland, an area of relatively high rainfall (1000mm) characterized by highly-leached infertile sands, another farmer plants traditional, open-pollinated and hybrid maize in the same field. The hybrids and open-pollinated varieties being early maturing are eaten either as green maize or are allowed to dry and are stored. The traditional varieties are not generally eaten green since they store better than modern varieties.

In both cases, it is the combination of traditional and modern varieties which allows for an extended period of harvest and offers the family greater food security. Each variety occupies an agroecological niche in the farmers' planting programmes.

In the same area a system of polyculture is practised whereby sorghum (*Sorghum bicolor* subsp. *caffrorum*) and millet (*Pennisetum americanum* subsp. *typhoideum*) are broadcast before the field is prepared by hand-hoe. Maize is then planted in holes one metre apart. Approximately 3 to 4 weeks later an indeterminate species of cowpeas (*Vigna unguiculata*) is planted between the maize with a similar spacing. Alternatively or additionally, pumpkins, cassava and groundnuts may be planted. Fields will be extended over time should rainfall prove satisfactory, which increases the period of harvest. The inclusion in the fields of naturally-occurring fruit trees such as marula *(Sclerocarya birrea)*, waterberry (*Syzgium cordatum*) and Natal mohogany (*Trichilia emetica*), constitute a form of agroforestry.

The utilization of wetlands, namely the swamp forests, for a variety of crops is another form of sustainable agriculture based on indigenous knowledge. Such systems are a form of swidden agriculture and are therefore sustainable only under low population pressure. The arbitrary and unilateral decision by the nature-conservation authorities (see CORD, 1992) in declaring such areas reserves has disrupted agricultural systems, threatened food security and jeopardized the prosperity of a considerable number of households. The indigenous knowledge of swamp farming handed from father to son is also threatened. The move to less favourable environments will test the experimental ability of farmers, as new innovations will be essential if yields are to be maintained.

Research challenges

While traditional systems are clearly innovative, adaptive and eclectic, equally self-evident is the lack of confidence many low resource farmers have in their own knowledge and abilities. On numerous occasions the Centre for Low Input Agricultural Research and Documentation (CLIARD) has been asked by farmers 'to show us how to farm'. But it is precisely this knowledge that must be sought, not as a platitude to make the farmer feel important, but in the realization that sustainable agriculture must be developed within a local context as it is out of this that traditional systems have evolved. At the same time it is important to understand that systems that were once sustainable may not continue indefinitely, and therefore, more often than not, it may be necessary to borrow from elsewhere.

But researchers must no longer see the transfer of technology from research institution to farmer as the solution. Solutions must come from farmers with researchers acting as process facilitators. This requires starting with the farmer determining problems, needs and priorities; searching with farmers for alternative solutions; developing hypotheses with farmers; testing and evaluating possible alternatives by farmers themselves and finally disseminating solutions to others.

An NGO's response

CLIARD was established as a specialist agricultural NGO to elicit the needs of farmers and their communities, and to work with farmers to research their problems and implement solutions. The NGO operates as a general consultant to various development committees, and as agricultural consultant to affiliated farmer groups who are its clients and to whom it is accountable.

A number of tools are utilized to research problems of farmers:

○ formal questionnaires with closed- and open-ended questions;
○ rapid rural appraisal techniques;
○ workshops with farmers;
○ farmers days and agricultural shows;
○ demonstrations on- and off-farm;
○ participant observation of farmers' practices.

The activities of CLIARD are built on the premise that without viable, representative and autonomous farmer organizations or rural institutions, agriculture for the low-resource farmer can never be sustainable. These organizations include small groups of women based on the cohesiveness of the neighbourhood group, farmer co-operatives and development committees at local, sub-regional, regional and national levels. Ultimately the formation of a rural social movement is perhaps the only hope for the rural poor to have their concerns addressed by government.

CLIARD works closely with farmers and their organizations, trying to increase agricultural production by offering an extension service backed by a team of agricultural specialists. Its links to other NGOs facilitate the implementation of an integrated strategy including water, sanitation and other development priorities. Knowing that its capacity will always be limited,

CLIARD has established links with the Universities of Zululand and Natal, state research institutions and the various departments of agriculture.

Changes in rural development policy are needed for farmers' needs to be met. This requires advocacy and a strategy of engagement with state structures. Rural organization building takes time and resources but many issues are pressing. Therefore, twenty-one NGOs and community-based organizations have established an umbrella body for advocacy and engagement.

Conclusion

As time passes, tradition changes and with it traditional knowledge systems and practice. But in this flux there is the danger that the old is discarded because it is old, rather than because it is no longer useful.

Man has moved from agriculture being part of a re-creation of the worldview where each season's planting was synonymous with the first-ever planting, to that of a mechanical process; and now to something incorporating aspects of old or new or perhaps something completely different, this we do not know. What we do know is that in changing ecosystems through agriculture, and thereby creating agro-ecosystems, we change both the equilibrium and stability of existing systems. The question, then, is how to create agro-ecosystems which mimic the ecosystems they have replaced yet at the same time offer productivity and prosperity to society.

In assuming that this process of change needs outside agencies to facilitate or perhaps initiate it, we run the risk, however, despite the best of intentions, of further alienating low-resource farmers from systems we still only partially understand. This remains our dilemma.

PART FOUR

CHALLENGING POLICY

Introduction

Kojo Amanor

The recent growth of interest in local knowledge reflects the fact that farming communities outside of agribusiness have been marginalized. While the process of marginalization has involved the downgrading of the knowledge of the farmer, it has an impact far beyond knowledge systems. Thus any approach which seeks to improve the farmers' lot or to empower them must go beyond promoting a positive image for local knowledge, to understanding the processes of marginalization and their implications for the life of the people.

The three papers in this section are concerned with policy dimensions, examining the implications of the commercial interests of agricultural technology development for the small farmer, and pointing to the necessity of change in policy at the international, national and local level in order to promote a climate which is favourable for local crop development. Opole and Cordeiro show how projects operating at a local level have attempted to deal with the marginalization of farmers in agricultural research. Attempts to take up issues related to marginalization has resulted in these projects questioning and challenging inappropriate policy frameworks within the national organization of research, which serve to foster the dominance of agricultural research institutions over the control of seeds. Mooney shows at the international level how the commercial interests of agricultural research have promoted legislative forms which threaten the ability of farmers to produce their own seeds, and how the contributions of farmers to the production of varieties has been downgraded, enabling commercial plant breeding firms to gain monopolistic control over seeds.

Opole describes the rationale behind a project which focuses on women's agricultural knowledge of indigenous vegetables. These crops have been marginalized by scientific research. Opole traces the factors which reinforce this marginalization within the ideology of a scientific tradition concerned with validating the products of its own research world according to abstract criteria, while downgrading the research traditions of the people. In the case of indigenous vegetables this takes the form of unfounded assertions about the nutritional inferiority and toxicity of the crops. This marginalization also finds ramification in other institutions which stress high input solutions and exotic crops through provision of credit incentives, and through the education system and media. She argues that science and technology development in Africa is divorced from people and has been influenced by ideological frameworks which have developed in narrow commodity sectors. It is only when the framework of research becomes concerned with incorporating producers' knowledge that sustainable food production systems can develop. Thus the basis for the development of a people-oriented science and participatory technology development involves an awareness and critique of the negative impacts of the commoditization of science and a revalidation of the knowledge systems of producers.

Cordeiro describes a project in south-east Brazil which aims to help farmers preserve and develop their own maize varieties, which are rapidly disappearing as a result of the aggressive promotion of hybrids. This involves the development of a participatory methodology through which farmers test the performance of different varieties and engage in breeding and crossing seeds. The project aims to build the confidence of farmers in their potential to develop their knowledge of seeds, a confidence which has been eroded by the expansion of commercial hybrid seed marketing. While the project has attempted to develop linkages with government agricultural agencies, the research programme has been threatened by the implication of new patenting laws which will enable seed companies to establish monopoly control over the breeding of local varieties. The processes of agricultural research are influenced and driven by commercial, social and political forces which are against the interests of small-scale farmers. It becomes necessary for farmers to organize to resist these socio-economic and political pressures.

Mooney is also concerned with the implications of commercial interests within agricultural research for small-scale farmers in the South. He argues that the knowledge of farmers in the south has made considerable contributions to international agricultural research, in the form of genetic materials and knowledge which have been distributed from international agricultural centres to commercial firms. But this contribution is both unacknowledged and not compensated for in this unequal exchange between the North and the South. This is further aggravated by patenting laws which enable firms to gain a monopoly over research, production and distribution of seeds which originate from farmers' varieties. As private companies move into the seed markets of the South, farmers are eventually going to pay for the end product of their own innovations. Thus a policy framework is needed which confronts these aspects of the increasing commercialization of research, and examines issues concerned with access to seeds.

These contributions show that commercial interests and development institutions frequently act to downgrade the creative capacities of farmers, and to reinforce industries' control over genetic materials and the marketing of seeds. They point to the importance of revalidating farmers' knowledge and innovatory abilities as a means of building local capacities for the conservation and development of genetic materials. For this to be achieved it becomes important to understand the institutional context of research and to develop a policy framework that is critical of the monopoly commercial interests of agricultural research which have resulted in the drive towards standardization, a drive which threatens farmers' creative abilities and the genetic diversity which has resulted from this creativity.

Revalidating women's knowledge on indigenous vegetables: implications for policy

*Monica Opole**

Within Kenya, women's local knowledge and the indigenous vegetables they have selected for cultivation have been downgraded by scientific knowledge. Scientific knowledge is more concerned with the products of its own research world and the validation of the abstract criteria it has developed. Assumptions of the scientific community on the poisonous qualities, poor nutritional content and difficulty of cultivating indigenous vegetables were proven to be unfounded by the research of the KENGO Indigenous Vegetable Project to be unfounded. The paper critically examines scientific research methods and conceptions, and the implications of the policy environments within agricultural sciences, the education system, and the media for local knowledge and indigenous crops.

The conventional view of development emphasizes the solution of rural problems by the application of scientific knowledge. Scientific agricultural knowledge today still operates within a 'top-down' or 'trickle down' approach where a few 'specialists' are custodians who control and own the scientific knowledge base and use intermediaries such as extension workers to pass on information to 'targeted' end-users. Local capacities and knowledge have not been recognized or incorporated into the mainstream of these development processes. People's search for self-sufficiency in food production thus takes place within parallel systems of production. There are certain systems requiring high external inputs of manpower, expertize, equipment and chemicals, and others utilizing local knowledge, manpower and biodegradable inputs. The difference between the two is that those requiring high external input are not sustainable. Low-input production systems, which depend on local knowledge, operate at low levels of productivity and have not received the same levels of research input to raise production to meet the demands of a growing population and market.

In the first system, modern scientists are seen to be operating in isolated spheres where they develop, test and produce products to be tried on targeted users. The methodologies and tools used very 'scientific', sometimes requiring

---

* Monica Opole dedicates her paper to farmer Janet Odindo, her mother and nurturer, who was instrumental in the development of many of the hypotheses now taken seriously by the formal sector. She also dedicates her paper to Dr Chweya and Professor Imungi, without whose support no self-respecting student agronomist and nutritionist would have taken this subject at post-graduate levels, and without whose input, policy would have not been so sensitized.

157

long and tedious data gathering and complex analytical methods before any subsequent trials. In this process, the end-user is the last to be consulted. The result is an accruement of scientific information and methodologies which, despite their interest for the scholar, are quite useless to the end-users, most of whom are illiterate.

Since the understanding of science itself requires the attainment of certain levels of literacy, following this system has a detrimental effect on end-users who cannot understand the scientific process, practices or results, leading to frustrations among extension workers and scientists who fail to comprehend why their scientifically-proven methodologies do not work.

In the second system, scientists first have to recognize that while rural or local people may not understand science *per se*, for technology to be successfully adopted, tools and techniques need to be developed together with the prospective target communities. From this perspective, it becomes easier for the researchers to work from the premise that useful local knowledge systems exist, though they may as yet be without a scientifically proven basis. Transfers of scientific knowledge may negate alternatives developed within local knowledge systems. In some instances, scientific knowledge has replaced local indigenous knowledge systems and practices and given rise to unintentional socio-cultural and economic consequences.

Improving traditional production systems requires a change of attitude regarding how local knowledge relates to science. This paper describes an approach developed by KENGO for working with women on indigenous vegetable crops within their indigenous knowledge base. It also examines the implication of agricultural and wider policy frameworks for those who work within indigenous systems of knowledge and resource use.

The Indigenous Vegetable Project: concepts, methods and objectives

The Indigenous Vegetable Project set out to develop a new form of relationship and interactions between scientists and end-users by developing innovative processes and ways of using and linking local knowledge to modern scientists. Certain assumptions and previously unchallenged myths about gender/power balance and ownership of knowledge had to be taken into consideration in the project's formulation and development. In the process, mistakes were made by scientists, custodians of local knowledge and users, but through these experiences the project has continued to evolve.

The concept of the Indigenous Vegetable Project grew out of the need to:

O widen and increase household food security;
O reinstate indigenous values and knowledge of marginalized crops;
O sensitize the public and thus open up a forum for deliberating and sharing information on indigenous value systems;
O document user preference in agronomic practices and nutritional values of indigenous vegetables;
O investigate women's knowledge and their power base in this area.

The project was designed to address the failure of conventional development models to bring about significant improvements in rural diets. This was seen as resulting from:

o a neglect and lack of respect for indigenous knowledge, preference and capacities of production and use:

o poor or low levels of understanding of the characteristics, processes and the power balance of ownership of the indigenous knowledge base;

o increased pressure by scientists on farmers to produce exotic or high yielding food crops so as to meet global projections at the expense of small-scale user-specific sustainable processes;

o pressure by the media through promoting lifestyles which support the use and production of exotic food crops rather than indigenous varieties.

In the Indigenous Vegetable Project, field agronomic trials have become an extension of the technological development process. Thus, instead of research and extension handling the problem-solving process, this now focuses on farmers, local knowledge systems, and synergic interactions for the achievement of desired effects for users, knowledge owners and scientists alike.

Methodology
A local knowledge system approach always tends to exceed the simple objectives of increased production and income set by most development projects. The tendency of scientists to focus only on outcomes and products rather than local knowledge in a problem-solving approach has been found in previous projects in Kenya to result in a misdirecting of well-intentioned assistance. The Indigenous Vegetable Project tried to avoid this pitfall by focusing more on the context of the extensive knowledge of people (women in particular). Rather than trying to solve a researcher-generated problem, based on abstract indicators of 'success', it tried to develop a multi-disciplinary approach to achieve a desirable 'process'.

The production and cultivation of indigenous vegetables
The indigenous knowledge base for production of semi-domesticated and domesticated indigenous vegetable varieties by women in Western Kenya covers all aspects of crop production and utilization.

The best site for a village home garden is an abandoned cattle *boma* (shed) which has not been used for a year. These sites are used for the production of species such as *Amaranthus lividus* (known as *Ombok akikra* among the Luo), favoured for its flavour, drought resistance and fast growth rate. Other sites for vegetable production in a village are home gardens outside the homestead (known in Luo as *Puoth alot* or *gunda*). Close proximity to the house ensures that varieties of indigenous vegetables stocked for seed production for the next season are available for consumption in the house. Within each household, seed production and storage are the reserve of women.

This system today is rapidly disappearing and being replaced by the modern kitchen with exotic vegetables, such as the *Brassica* species commonly known as Sukuma wiki ('push the week'), tomatoes and onions, which are promoted by agricultural extensionists. There is a danger of high intake of chemical pesticides where there is intercropping of exotic and indigenous vegetables as farmers lack the knowledge of the time lapse required before harvesting when using pesticides on farms.

Preparation and planting

A previously uncultivated site is either ploughed using oxen (if it is for commercial production) or a big *jembe* (hoe) and harrowed to a fine tilth using a vegetable hoe or a hoeing stick, as done by the Akamba community of Kenya. A mixture of seeds is either broadcast, as in the case of *orundu* or the home garden. In a *shamba* or commercial indigenous vegetable garden, seeds are planted in rows at approximately 30cm intervals between rows. The rows are prepared using a hoeing stick and seeds mixed with soil or sand in a ratio of 1:2 and drilled in a line for commercial production. In such a system more than two varieties may be planted together. For example, *Amaranth* may be intercropped together with *Corchorus olitorius* and *Vigna nguiguculata*. Row planting in the traditional system uses the same method for intercropping with millet and sorghum and, more recently, maize.

The women normally consider seasonal weather patterns before selecting the species for intercropping with cereal crops. For example, *Corchorus olitonius* will yield less bitter leaves when planted with the new moon rather than a waning moon. Local knowledge also recommends planting during light rains occurring either before or just after the heavy rains for higher leaf yield. This practice also protects the plant from mildew, which is at its peak during days of heavy rain.

Building linkages with local knowledge

Modern science often makes assertions about local knowledge systems and practise, which are not rooted in substantial analysis but reflect bias and gaps in research. In the instance of indigenous vegetables in Kenya, orthodox practitioners and scientists held a common belief that indigenous vegetables cannot be cultivated, especially the wild varieties of some species of *Amaranthus* and *Vigna*, that they were poisonous, and that they were nutritionally inferior to exotic cultivars. This resulted in a research gap between local and scientific knowledge concerning the premise for initiating research, the choice of species or varieties for incorporation in development programmes and farming systems, and the basis for starting participatory agronomic work.

Using participatory research methods, twelve varieties of indigenous vegetables were selected for co-cultural agronomic trials and nutritional analysis. The species were selected on the basis of local knowledge, user taste preference, acceptability, level of utilization, commercial viability, and nutritional and medicinal use.

To make the initial link between local knowledge and scientists, it was first necessary to demystify negative myths such as the nutritional inferiority of indigenous vegetables. Initial analysis was done on fresh samples collected from farms for both nutrients and anti-nutrients. This was followed by more detailed analysis and ranking of nutritional value. Table 4.1 shows levels of nutrients in indigenous and exotic vegetables species selected for detailed work.

The other myth which needed to be laid to rest was that indigenous vegetables are poisonous. This belief stemmed from the observation that indigenous vegetables such as *Solanum nigrum* and *Gynandropsis gynandra* are bitter. Modern scientific knowledge points out that the main anti-nutrients commonly found in indigenous leafy vegetables are nitrates and tannins. High level consumption of nitrates results in both acute and chronic toxicity, especially in small children. Tannins, on the other hand, compete for protein during the ingestion process, resulting in deficiency syndrome even in diets containing sufficient protein. Research revealed that the levels of these anti-nutrients in both *Solanum nigrum* and *Amaranthus* species were not high enough to warrant any concern (Table 4.2).

Local knowledge stresses the use of farmyard manure rather than hemical fertilizers for food production. Some species of indigenous vegetables such as *Solanum nigrum* produce fewer leaves and show poor growth rates when planted with chemical fertilizers. Long-term use of such inorganic fertilizers also results in the increase of aphid attack and root worms on *Solanum nigrum*. Women have also reported that the species grown using organic fertilizers are more bitter

Table 4.1: Nutrient contents of some Kenyan indigenous and exotic vegetables (per 100g edible portion)

Vegetables	Vitamin A (mg carotene)	Vitamin C (mg)	Protein (g)	Calcium (mg)	Iron (mg)	Raten (g)
Indigenous						
Gynandropsis	6.7–18.9	127–177	5.4–7.7	434	11.0	1
gynandra						6
Solanum nigrum	2.7–7.9	37–141	3.2–4.6	215	4.2	4
Amaranthus spp.	5.3–8.7	92–159	4.0–4.3	800	4.1	3
Croalaria brevidens	2.9–8.7	115–129	4.2–4.9	270	3.8	2
Corchorus olitorius						
	3.9–5.4	170–204	4.5–5.5	270	7.7	
Exotic						
Curcurbita spp.						
Brassica carinata	2.4–5.3	170–172	3.2–4.2	40	2.1	2
Spinacia oleracea	3.7–5.7	102–142	3.6–3.8	520	6.0	5
Brassica oleracea	2.8–7.4	1–59	2.3–3.1	60–595	0.8–4.5	6
var. capitata	Tr.–4.8	20–220	1.4–3.3	30–204	0.5–1.9	7
Brassica oleracea						
Lactuca sativa	Tr.–0.4	8–114	1.8–3.4	13–43	0.2–1.9	
	0.15–7.8	3–33	0.8–1.6	17–107	0.5–4.0	

Source: M.Opole, J.Chweya and J.Imungi (1991)

Table 4.2: Effect of chemical fertilizers and farmyard manure application on the nutrient contents of *Solanum nigrum*

Level of manure or fertilizer application (kg/ha)	Dry matter (%)	Protein (%)	Total Ash (%)	Beta- Carotene	Vitamin C (mg/100g)	Nitrates (mg/ 100g)	Tannin (mg/ 100g)
Control	18.2	5.8	2.8	4.8	125.0	400.0	42.3
100kg DAP	12.8	4.3	2.0	3.6	76.0	346.0	50.0
200kg DAP	13.6	4.4	2.2	4.8	121.0	122.0	43.0
100kg DSP/100kg CAN	13.3	4.1	2.0	5.2	99.0	359.0	40.3
100kg DSP/200kg CAN	11.5	3.5	1.9	4.9	107.0	359.0	45.0
200kg DSP/100kg CAN	11.7	3.8	1.8	2.8	89.0	304.0	40.6
200kg DSP/200kg CAN	11.8	4.0	1.9	5.9	112.0	266.0	45.8
5T Farmyard manure	11.7	3.8	1.8	7.5	85.0	283.0	51.8
10T Farmyard manure	15.2	5.4	2.2	5.0	139.0	228.0	49.0
20T Farmyard manure	14.7	4.7	2.4	4.8	120.0	176.0	39.7

tasting. In agronomic trials it was found that total leaf-top yields increased with the application of farmyard manure rather than chemical fertilizers.

Policy implications

The shift in the use of knowledge systems from 'indigenous' to 'exotic' or 'modern science' in food production cannot fully meet local household food needs and security. New cultivars and crops such as cabbage, which require inputs of chemical fertilizers, pesticides, insecticides and herbicides, have immense bearing on local diets and food security particularly when considering the difficulties of sustaining exotic cultivars under local knowledge and ecological conditions. All over Kenya, field reports show that in the event of poor rainfall or massive pest attacks new crops fail because of the tendency to cultivate them in monoculture. Producers and consumers both incur great losses in a food production system which stresses the use of high inputs, as compared to local production systems.

The project found that in bridging the gap between agricultural scientists and custodians of local knowledge not only is a change in individual attitudes required, but policy changes also need to be made in the national planning process to facilitate the participation of local knowledge systems in current mainstream frameworks of development. Taking the case of modern agricultural inputs in Kenya, one finds that high potential areas, which have traditionally benefited from high inputs of chemically-based pest management, credit and finance, enjoy more privileges than the low potential areas which tend to rely more on indigenous knowledge and practices in crop production and protection.

Although producer knowledge in this area has withstood the test of time, changing circumstances make it vital that small producers have access to 'improved' knowledge of the environment and crop protection so as to increase subsistence levels and produce surplus for sale.

While the conceptual framework and methodology used by the project could address issues concerned with the validity of women's knowledge in agricultural production, other factors with important implications for the indigenous knowledge base, including the ideological underpinnings of the educational system and media system, need to be addressed.

Women's knowledge

The production of indigenous vegetables rests solely on the immense time-tested knowledge power-base of women in agronomy, nutrition and post-harvest practices. By taking into consideration this resource in setting up its conceptual framework, the Indigenous Vegetable Project has avoided the tendency of modern agricultural processes to weaken women's already precarious power base. By using participatory research methods, the development of the local knowledge base on nutrition and agronomy has been strengthened and 'modern' myths and beliefs about the nutritional inferiority and poisonous qualities of indigenous vegetable crops have been undermined. Such an approach has given a synergism not only to local knowledge but to scientists who find it easier to conduct research relevant to future world food needs.

Education

Modern school education is a pre-requisite to understanding modern science. In contrast, informal education through the use of folk-media, concepts and phenomena has been used to synthesize local knowledge over long periods of time.

Education in the current system of development has contributed to the gradual loss of indigenous knowledge by separating young girls from their mothers as they go to school. In school, curriculum subjects relating to agriculture and nutrition focus more on the methods of production and processing and on exotic food varieties. From kindergarten school, children are introduced to exotic foods, as illustrated in the letters of the alphabet, where 'A is for Apple' and 'C for Carrot'. Most rural children, particularly those in low potential areas, will never see an apple in their lives.

At primary and secondary level, most agricultural subjects and agriculture clubs show the same pattern of promoting exotic foods. Carrots, cabbages and tomatoes are the common species grown by young prospective leaders to depict a typical 'kitchen garden'. The university curriculum also focuses on export-production knowledge. But a uniform food supermarket where only a few selected species are available for consumption, critically endangers the roots of sustainable existence in the faced of disasters like a new disease or changed weather patterns.

The media

As a tool for communication, the media has also contributed immensely to the rapid loss of knowledge of local production systems. Large multinationals have used the media to promote their chemical products and inputs at the expense of traditional foods, thus contributing to the marginalization of indigenous knowledge systems of production.

Advertisements, showing idealized crop qualities such as yield and colour, rather than taste and user preference, have been recognized within the Indigenous Vegetable Project as one of the major factors causing farmers in high potential areas to forego production of traditional food crops alongside 'modern' industrial crops such as tea, coffee and sugar-cane. Equating glossy production systems to cash has resulted in malnutrition and inadequate food production systems in ecological zones which were previously the bread-basket (or *ugali*-basket) of the continent. This trend has widened the gap between local knowledge and scientists.

Redefining research
Science exists in two forms: there are indigenous or local knowledge systems which are more people oriented, and modern scientific knowledge which has many disciplinary sub-systems and is technology oriented. Science and technology development in Africa has been influenced by external ideological models generated for developing countries. But for sustainable development to take place, specifically in food production, African policymakers must be able to differentiate between ideological and practical frameworks of technological development at all levels. Methods must be found to 'move' women's indigenous knowledge beyond the existing ideological framework to more practical levels, particularly in local food production systems as illustrated in this paper with the example of the Indigenous Vegetable Project.

It is only by elevating producer knowledge into the current framework of development that sustainable food-production systems can be maintained. As realized at each stage of this project, people still need a better understanding of the systems within which they exist. Focusing on high technological research only seems to compound the problem of improving basic lifestyles. It is only when attitudes to the use of modern science for research are changed from being *laboratory* to *people-oriented* that we will be able to develop a holistic approach to true agricultural and local knowledge advancement.

Rediscovering local varieties of maize: challenging seed policy in Brazil

*Angela Cordeiro**

The paper describes the experiences of a programme run in six districts of southern and southeastern Brazil by PTA, a network of NGOs which aims to reintroduce local varieties which have been eroded with heavy promotion of green revolution technologies. The problems with hybrid seeds are described in the context of escalating prices for commercial seeds and inputs and lower prices for farm produce. The paper describes a participatory strategy which involves farmers in collecting seeds, evaluating their performance through on-farm trials, breeding and crossing seeds, and multiplying selected seeds. The programme involves collaboration between farmers, NGOs and government agencies. The paper raises the implications of new patenting laws, and their ramifications for collaboration between NGOs and government agencies in genetic resource conservation.

Since the early 1960s the process of agricultural modernization in Brazil has marginalized the majority of farmers, deepening existing social inequalities. The dissemination of green revolution technologies has brought about serious social, economic and environmental problems. In response to this, the Alternative Technologies Project (PTA), a network of nineteen non-governmental organizations working within twelve Brazilian states, has developed actions with small-scale farmer organizations to search for alternative models of agricultural development.

An important focus of this work has been developing initiatives which encourage farmers to build self-reliant capacities to produce their own seeds. Since 1990, the PTA Network has developed a programme with farmer organizations in the south and southeast regions of Brazil, which seeks to rescue and reintroduce farmers' varieties of maize in place of hybrid brands. Major concerns arising from this experience include the possibilities of using and conserving local plant genetic resources; the development of methodologies for farmer participation in the generation of technology; the needs for policy changes in relation to the promotion of hybrids and marginalization of local varieties; and

---

* Angela Cordeiro dedicates her paper to Mr and Mrs Nunes and all farms families which still maintain their local varieties and support her work with their knowledge and their seeds.

the implications of patenting laws for farmer experimentation with crop varieties and the development of research linkages.[1]

Losing local varieties

Maize is one of the most popular crops in Brazil. In the 1980s the area planted with maize totalled an annual average of 12.4 million hectares, with an average annual production of 22.7 million tonnes. Small farmers are responsible for 53 per cent of the national production and many of them rely on maize as their main source of income.

Maize occupies a prominent place in national agronomic research. Brazil was one of the first tropical countries to produce and distribute hybrid maize, and research with hybrids dates back to the 1930s. From the end of the 1960s a programme was launched which gave farmers incentives to substitute local varieties with commercial hybrids. This was reinforced by a strong propaganda campaign which classified as 'backward' all the farmers who did not adopt the 'modern' hybrid seeds. In the universities, technicians were taught that only hybrids guaranteed good productivity and this information was propagated through the extension services. To encourage the adoption of modern practices the government handed out cheap and easy credit terms.

Given these pressures and incentives, commercial varieties were adopted on a large scale in the main maize producing areas and local varieties disappeared. In these areas, there is now at least one generation of farmers who do not have any experience of planting local varieties and producing their own seeds. Local varieties are largely confined to the memories of their fathers and grandfathers.

Problems of dependency on hybrid seeds

Agriculture in Brazil has been greatly affected by the economic crisis of the eighties. Agricultural credit became more expensive and less accessible for small farmers. The cost of purchasing inputs has increased but prices for agricultural produce have declined. The increased costs were not only reflected in chemical inputs such as fertilizer but also in hybrid seeds. Many farmers stopped purchasing commercial seeds and started to plant second, third and even fourth generations of hybrids. Combined with problems of soil fertility, this has resulted in low productivity, further eroding incomes obtained from maize.

Given the importance of maize within the production system of small-scale farmers, NGOs within the PTA network began to discuss with farmers

---

[1] The following NGOs are taking part in this programme: PTA Association of Alternative Technology Programmes; AS-PTA Consultants in Alternative Agricultural Projects, ASSESOAR Association for Rural Studies, Orientation and Assistance; AVICITECT Vianei Association for Cooperation and Exchange in Labour, Education, Culture and Health; CAA-NM Alternative Agriculture Centre for Northern Minas; CAR Ecological Agriculture Centre of IPE; CAPA Support Centre to Small Farmers; CAT Centre of Technical Advice; CETAP Centre for Popular Alternative Technologies; CTA-ZM Alternative Technology Centre for Zona da Mata; REDE Network Association for the Exchange of Alternative Technologies.

166

alternatives for improving the system of maize cultivation with minimal production costs. In the course of searching for alternatives, some farmers were found who had maintained their local varieties and were obtaining satisfactory yields under not very favourable production conditions. This went against the myth of the hybrids and attracted attention.

A series of training courses were organized for PTA technicians to discuss constraints and potentials of promoting and disseminating local varieties. In 1990 the PTA network defined a common strategy for work in the southern and southeastern regions of Brazil, where pressures for the modernization of agriculture and adoption of hybrids were more intense. The valuation of local varieties and farmer participation in research were identified as the main strategies to be followed in the quest for solutions to dependency on hybrid seeds.

It was difficulty to find a source of advice and expertise on local varieties within the formal plant breeding sector, since the majority of breeders were dedicated to working with hybrids and did not encourage the use of 'primitive' varieties. However the PTA network sought to develop linkages with the formal sector both to develop a more systematic approach to local maize varieties and to exercise an influence on public sector institutions, encouraging them to meet the needs of small farmers.

A strategy for recovering local varieties

A research strategy was designed comprising four stages, each subdivided into a series of methodological steps (Figure 4.1). The first stage consists of the rescue of local varieties and relevant cultural information and values associated with them. After mapping out the varieties maintained by farmers within a specific locality, seed samples are obtained for multiplication by farmer groups. In order to prevent the possibility of eventual loss of the varieties, a portion of the samples collected is stored in a community seed bank and in one of the banks of the NGOs of the PTA Network. The main principle behind this is not to set up a collection of germplasm, but to maintain a reserve of seeds in good storage condition in the transition period between the rescue and widespread use of the varieties.

The objective of the evaluation stage is to check the performance of varieties which farmers are unfamiliar with in their localities. This is achieved through comparative varietal trials under farmers' cropping conditions. One of the trials, the National Trial on Native Maize (NTNM) was conducted at the network level in co-operation with the National Centre for Research on Soil Biology, which is connected to the Brazilian Company for Agricultural Research (EMBRAPA). The NTNM evaluates local varieties rescued in the south and southeast regions. Trials are established on parcels measuring $10m^2$ under highly controlled conditions. The same trial is replicated in many different localities by groups of participating farmers. Other experiments are also carried out at the local level in accordance with the interest of farmers.

In the process of evaluation, some varieties reveal characteristics which farmers wish to eliminate or improve upon. This is achieved through selection or crosses between varieties. With more information on varieties and their cultural requirements, farmers can develop further options and begin production of seeds at the community group level or individually. These research procedures do not follow a linear progression since many stages occur simultaneously.

Lessons from on-farm seed selection

NTNM has made a significant contribution to local varietal seed culture. The results obtained in two years of experimentation have been promising, confirming the potentials of local varieties. In 1990, the first year of NTNM, trials were effected in 12 localities in which 49 varieties were evaluated, including 28 local varieties, 4 hybrids produced by farmers, 14 varieties from EMBRAPA and 3 commercial hybrids as controls. Results of agronomic trials have shown that local varieties perform a little behind the hybrids. In 1991, trials were carried out in 18 localities in which 31 local varieties, four farmer-developed hybrids, 12 EMBRAPA varieties and two commercial hybrids were evaluated. The results show the same tendency as the previous year. In 1992, recently-rescued varieties are being incorporated into trials held in 21 localities.

NTNM has its limitations in relation to the participation of farmers in research. It has been developed together with a public agricultural research organization. As a result, procedures tend to be formalized and follow pre-established rules which allow little scope for farmers to define the ways in which trials should be conducted. Many of the participants question the validity of the experimental procedures used and doubt that results obtained on small plots under highly-controlled conditions have relevance for their farms. These limitations are being compensated for by the local trials where farmers participate more in defining the design and evaluation parameters and procedures. They usually prefer to do their experiments in larger areas without replication, in conditions more representative of their cropping conditions. This suggests that one should not be a slave of statistical methodologies. The search for genuinely participatory procedures implies finding ways of supporting experiments obtained in conditions quite different from those determined by formal research.

Farmers have their own breeding objectives and parameters. These are not often taken into consideration by plant breeding programmes, which are remote from the farmers. The PTA programme is attempting to develop new ways of carrying out plant breeding, which take into consideration the diversity of environments, farming systems and cultural values. It attempts to decentralize plant breeding and promote farmer participation. The main role of the technician is to disseminate information to farmers which can help them conserve and improve their traditional practices.

Local seed production is more feasible when it is developed at the group level. Maize is a cross-pollinated plant and its seed production requires special care to avoid contamination by other varieties and hybrids. Farmers often have small plots and cannot isolate areas for seed production. Farmers experimenting in the PTA programme have found that the solution to this problem is to produce seeds at the community level. Farmers with the largest areas of land lend plots for maize seed production by the community group. The work is developed collectively and the harvest is shared among all participants according to the time spent on seed production activities.

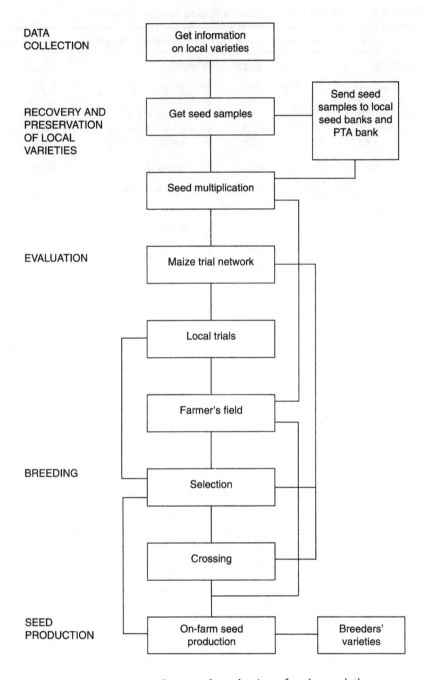

Figure 4.1: *Strategy for on-farm seed production of maize varieties*

Table 4.3: National trial of native maize — yield (kg/ha)

RESULTS 1990/91(*)		RESULTS 1991/92(**)	
HYBRIDS	**YIELD**	**HYBRIDS**	**YIELD**
1 BR 201	6985	1 AG35 x C 511	5077
2 XL 560	6482	2 BR 201	5019
3 XL 678	6238	3 AG 28 x C 408	4791
4 AG 35 x C 511	6080	4 BRASCHUNHA	4656
5 AG 28 x C 408	6079	5 AG 64 x P 307	4622
Average	**6372**	**Average**	**4833**
BREEDER'S VARIETIES	**YIELD**	**BREEDER'S VARIETIES**	**YIELD**
1 BR 106	6420	1 CMS 39	5036
2 CMS 39	6068	2 NITRODENTE	4545
3 EMPASC 151	5820	3 COMPOSTO VEGA	4426
4 NITRODENTE	5758	4 SINTETICO ELITE	4320
5 BR 105	5661	5 BR 105	4212
Average	**5945**	**Average**	**4512**
LOCAL VARIETIES	**YIELD**	**LOCAL VARIETIES**	**YIELD**
1 VARGEM DOURADA	5897	1 CAIANO SOBRALIA	4690
2 AMARELAO	5578	2 AMARELO PAULISTA	4424
3 PALHA ROXA B.E.	5128	3 CATETO ALEGRE	4245
4 ASTECA	5065	4 COMPOSTO MINEIRO	4198
5 MAIA ANTIGO	5055	5 ASTECA (ZM)	4170
Average	**5336**	**Average**	**4345**

*average of seven experiments   ** average of eighteen experiments

Conservation of local varieties and patenting laws

This experience in the conservation of local varieties is very recent and many questions still remain unresolved. A few years are needed before we can know the scale and rate at which local varieties are reoccupying their former place. Time is also needed to assess the extent to which a consciousness is developing among farmers of the importance of using and conserving local genetic resources, in regions in which hybrids have been heavily promoted to the detriment of local varieties and farmers' self-confidence in their skills in seed selection and multiplication. There is a tendency on the part of farmers to search for the shortest road, opting immediately to distribute massively EMBRAPA varieties which come 'ready off the shelf' in their communities. Thus the rescuing of varieties also involves the rescue of knowledge which has been lost with the seeds. This requires time. Farmers, who have been subjected to the aggressive propaganda of the green revolution, will not recover their self-confidence overnight. It is not easy for them to believe that it is possible to develop local varieties which will perform as well or better than those which are packaged on the shelves of agricultural supply stores.

Despite the short duration of the programme, preliminary results are encouraging. The myth of the superiority of hybrids no longer occupies its previous dominance and farmers have developed more trust in local varieties. Presently there are some 1000 families distributed in six Brazilian states directly involved in this work. In the last harvest season, the groups produced 50 tons of their own seed. The economic crisis has made the production of local maize varietal seeds highly attractive and every day there are new groups interested in joining the programme. Care is needed to guarantee that there is also an expansion in the quality of the programme. It is important that the information and values associated with the seeds are also disseminated with the local varieties.

The biggest challenge to the continuity of this work does not relate to technical and operational problems, but to new patenting laws for life forms waiting to be approved by the Brazilian Congress in 1993. This has implications for farmers' seed production programmes. If the law is approved it will enable seed companies to establish monopoly control over the breeding of local varieties. This may lead to farmers being prohibited from using their own varieties of beans, rice, maize, wheat, etc. Farmers cultivating other crops may face the same problems as maize farmers, including dependence and the high cost of seeds.

Should this happen, the relationship between the PTA network and government agricultural research institutions will become more complex. It will be impossible to maintain a cooperative relationship involving access to the germplasm maintained by farmers under a regime of patent protection. The seed programmes developed by the PTA network could also be considered illegal. The adoption of patents over life forms may compromise initiatives for the conservation and utilization of plant genetic resources.

Since 1991, when this law was sent to the Brazilian Congress, its implications have been discussed by farmers. Farmers have suggested that if the law is approved they will not observe it but engage in civil disobedience. They argue that producing their own seeds is a basic human right and that nobody, governments, companies or GATT, can take it over. The farmers and the PTA Network will have to reorganize autonomously to continue this work and be prepared to go beyond the legal questions which may arise and challenge the patenting laws.

Exploiting local knowledge: international policy implications

Pat Roy Mooney*

While local knowledge is presently very much in vogue among agricultural and development researchers, farmers in the South have already made a great contribution to the plant genetic resources of the North, through plant accessions in international agricultural centres. The paper draws attention to this through drawing up estimates of the value of the transfers of plant genetic materials to the North. But these transfers are largely unacknowledged and have been freely utilized by a commercial system which seeks to limit, protect and patent the use of genetic materials. The paper examines concepts of trusteeship, benefit and ownership within the international system of agricultural research, and the implications for farmers in the South and for the commoditization of research.

In recent years local knowledge and traditional farming technologies have become a growing fad and serious factor for institutionalized agricultural research. Indigenous knowledge became a tenet of UNCED's Agenda 21, recognized as an important base for building sustainable development and as an input for biotechnology development. Recognition is important in that it provides technical, institutional and financial support for local knowledge. Behind this interest in local knowledge, however, are also more disturbing preoccupations which threaten community interests. This paper considers some of these issues at the international level.

The interest in farmers' knowledge coincides with a period of the expansion of interest in biotechnology and international manoeuvring to gain control over this industry at the expense of farmers. At the national level, governments are also introducing policies to cut back on agricultural research and slash extension services. While NGOs and community development is being promoted, this may hide the long-term impoverishment of agricultural research and its implications for the interests of the farmer. Thus the present interest in small-scale farmers must be viewed within the larger framework of international agricultural and agri-business policies.

Genetic resource policies: trust and benefit

Terms such as 'farmers' rights' and 'informal innovation systems' first arose in the context of a debate over genetic resource control within the United Nations. Although substantial progress has been made over the years, particularly at FAO,

---

* Pat Mooney dedicates his paper to Richardson Thomas, a St. Lucian farmer and banana grower he met in 1967.

many issues remain outstanding or, where resolved, have yet to find a practical implementation. Many of these issues relate to farmers' rights and involve the wider international agricultural research community. They include concerns about trusteeship, ownership and who benefits from the value of the genetic materials of farmers in the South.

Trusteeship

The Rural Advancement Foundation International (RAFI) is in the early phases of an evaluation of the trusteeship role of the Consultative Group on International Agricultural Research (CGIAR) and its International Agricultural Research Centres (IARCs) with respect to their *ex situ* collections of genetic material. The trend in policy formulation within the CGIAR is to argue that each centre holds genebank materials 'in trust' on behalf of farmers in developing countries. This raises a number of problems and difficult questions:

o Is the legal status of each genebank clearly defined? Most IARC arrangements suggest that genebanks are an asset of the IARC and, therefore, ultimately the property of the country hosting the IARC.

o If the genebanks are not assets but 'trusts' who entrusted the CGIAR with this responsibility and what are the conditions of the trust? The IARCs are currently negotiating with the FAO to be given retroactive 'trust' over their collections. This is a convoluted negotiation since the IARCs seem to want unconditional trust.

o How is genebank material and research products delineated? Do IARCs have the rights to claim proprietary ownership over products based upon genebank materials or is there also a trusteeship element in their ownership?

o Can IARCs patent genebank material or trade it as part of their intellectual property. And can they enter into transfer of technology contracts with private companies?

Most of the accessions in genebanks are so-called 'landraces' (folkseeds, or farmers' varieties are better terms). These accessions form part of the intellectual integrity of farming communities. Germ-plasm is a genetic resource whose importance to conventional plant breeding and biotechnology increases daily. Genetic resources are often incorrectly described as the 'raw materials' for biotechnology. They are in reality the products of the intellectual contribution of informal innovators. But before they can be appropriated by private companies, the intellectual content of these resources has to be demeaned.

Ownership

IARC thinking on the parameters of intellectual property rights is ambiguous. We do not know of any instances in which IARCs have expropriated genetic materials held in trust for their own commercial advantage. But Hope Shand of RAFI has identified some recent instances in which IARCs have overstepped their self-appointed roles as trustees and used farmers' genetic materials as bargaining chips in negotiations which are resulting in the extension of patents. This includes the following five examples:

o Rice research financed by the Rockefeller Foundation, involving the participation of the International Rice Research Institute (IRRI) and Cornell University, has led to Cornell applying for a series of patents and offering to sell access of information to biotechnological corporations in the USA.

o In 1991, the International Potato Center (CIP), in Peru, signed a contract with Plant Genetics Systems (PGS) of Belgium to trade genebank material for access to a transgenic potato resistant to potato tuber moth. PGS has exclusive rights to the germ-plasm in industrialized countries and CIP has the right to the use of the material in the South. As a consequence of this agreement, CIP is obliged to refuse requests for this germ-plasm from other organizations based in the North.

o Frito-Lay, a Pepsico subsidy, was allowed access to CIP genebank accessions, to screen for germ-plasm with potato-chip processing qualities. Frito-Lay took useful germ-plasm samples back to the USA. It is now developing proprietary varieties which can be patented and then marketed in such countries as Korea and Taiwan, where Frito-Lay has large operations. CIP traded access to the genebank for information from Frito-Lay's screening documentation.

o EscaGenetics, another large biotechnology corporation, has also obtained germ-plasm from CIP which is currently developing into patentable material. EscaGenetics is testing its potatoes in a number of developing countries, including Egypt.

o The International Center for Tropical Agriculture (CIAT), in Colombia, is negotiating intellectual property rights over two new bean varieties with a French public sector institution. Royalties will be disposed of by CIAT. Officials have conceded that one of the varieties is heavily based on a Chilean accession in their genebank and have deliberated if they should turn the profits over to Chile.

CIP officials have argued that their genebank resource is their 'comparative advantage' in negotiating technology transfers with private companies. However this clause does not exist in any formulation of 'trust' agreements with farmers. Neither have governments from whose sovereign lands the germ-plasm originated been consulted.

There are serious implications for farmers in collaborating as informal innovators with institutes which accept the possibility of patent monopolies over genetic resources. Recognition of the validity of the innovations of the informal system may lead to the expropriation of both farmers' varieties and farmers' knowledge about these varieties. This raises key issues regarding how this appropriation of genetic resources can be prevented, and how farmers can be compensated for the use of the products of their innovation and work. These issues also apply to national agricultural research services and universities which are also becoming interested in patenting living materials.

It is important to understand that in a number of industrialized and Third World countries it is possible to obtain exclusive patent rights over almost everything. Products, processes and parts of life are all patentable in countries like the USA. Recently, British and American public sector health institutes applied for more than 4500 patents on genes and DNA fragments drawn from human brain tissue. Although the US patent applications have been rejected, the dispute will probably go to the US Supreme Court. If they are successful, the US National Institute of Health and the British Medical Research Council will have overturned two of the main tenets of patent laws. The patent claimants concede that they are using automated invention systems which simply discover previously unrecorded genetic material and claim rights to patents over them. Their approaches involve no inventive steps and they do not know the utility of their discoveries. This move has already damaged international research efforts on the maize and rice genomes. If the two health institutes win, intellectual property rights will be obtainable without any requirements of an intellectual capacity to undertake innovative research. 'Inventors' will only need a visa to patent the varieties of Third World farmers and the materials of the tropical forests.

Benefits to the North of farmers' genetic resources

It is difficult to quantify the intellectual contributions of Third World farmers to industrialized countries' genetic accessions. Most genebank directors in the North privately acknowledge that farmers' varieties have made an immense contribution. A large proportion of commercially- usable genetic materials transferred to the North pass through IARCs, either directly through their genebanks or as 'improved' material for field trials. RAFI is attempting to establish realistic calculations on the value of farmers' varieties exported in this way.

The best information is available for wheat material, obtained by the North through the International Maize and Wheat Improvement Centre (CIMMYT) in Mexico. At RAFI, our preliminary estimates are based upon information from only four industrialized countries: the US, Italy, Australia and New Zealand. For the US, our estimates are based upon a 1982 OECD report which, somewhat cryptically, estimates the value of Third World germ-plasm to the USA at $500 million per annum. An internal 1983 study by Canadian and US government wheat breeders came up with a similar figure specifically related to CIMMYT germ-plasm. RAFI believes these figures are conservative estimates. Two studies by Dana Dalrymple (1986) of USAID reveal that 21 per cent of the entire US wheat crop was sown to semi-dwarf material derived from CIMMYT. RAFI estimates the value of the CIMMYT contribution at roughly US$1.8 billion in the mid 1980s.

In a report for the Crawford Fund for International Agricultural Research in Australia, Derek Tribe (1991) offered different calculations for the value of CIMMYT material to Australia and New Zealand. For Australia two different data sets for calculations were presented: a low estimate of US$75 million per annum, and the high figure (based on RAFI's extrapolation of other data cited by Tribe) of approximately US$122 million per annum. Tribe also cites a 1990 estimate for CIMMYT's contribution to New Zealand of NZ$338,000 (US$0.3 million) per annum.

A study from INTAGRES (1992), a CGIAR documentation and information centre based in Rome, concludes that CIMMYT's annual contribution to the Italian durum wheat crop was not less than $300 million.

Thus the total of CIMMYT wheat germ-plasm to these four industrialized countries is not less than $875 million per annum, according to official estimates.

This data is summarized in Table 4.4. RAFI has taken each of the estimates for CIMMYT's contribution to the four countries and extrapolated them for all industrialized countries on the basis of the particular country's percentage of total wheat production in the North. The calculations on the basis of one set of figures are also extended to the other countries. Thus estimates for the US, which averaged 18 per cent of the North's total wheat production over the period 1986-1990, are calculated on the basis of the OECD report, on RAFI's extrapolation on USAID acreage data in 1986, on Tribe's figures for Australia and New Zealand, and on INTEGRES estimates for Italy. Each country is measured in the same way. The result is a range (see row 'North'), estimating the total annual value of CIMMYT's material to the North, running from less than US$300 million (based on the New Zealand rate) to more than US$11.5 billion (based on the INTAGRES Italian rates). The extrapolation of the OECD data and the high estimates of Tribe's data gives a figure of about US$2.7 billion per annum. RAFI believes this to be a reasonable estimate of CIMMYT's real value to the North. Given that CIMMYT's budget in 1990 was US$27.1 million, the North's returns for investment in CIMMYT are one hundred fold.

RAFI has also attempted to estimate the value of IRRI and CIAT rice materials, CIAT beans, and CIMMYT maize on the basis of annual production in the North. These figures must be treated with caution, and only offer crude estimates of the contribution of informal innovators. They do, however, illustrate that in the final analysis the contribution of Third World farmers to the North is enormous.

A USAID study (Dalrymple, 1986) showed that 73 per cent of the semi-dwarf rice acreage in the US was based on IRRI material. Semi-dwarf varieties contributed to 22 per cent of the entire US rice crop. Extrapolating this data, RAFI estimates that the annual farmgate contribution made by IRRI amounted to about US$176 million in 1984. The semi-dwarf share of the American rice harvest has continued to grow, but RAFI has kept the figure at the 1984 level, indicating a conservative estimate. Since the US crop equals about 26 per cent of the North's total rice production, the total value of IRRI material to the North is estimated to be about US$677 million per annum. In 1990, IRRI's budget was US$30.6 million, which offered the North a 22 fold return on investment per annum.

A US study of maize (Goodman, 1985) estimates that in 1985 only one-tenth of one percent of the value of the US maize crop was based on 'tropical' exotic germ-plasm. In the mid-1980s this tiny percentage was worth US$20 million per annum. Since US maize is estimated at about 68 per cent of total maize production in the North, the total value of Third World germ-plasm to the North is only about US$29 million. If all this material was derived from CIMMYT this

Table 4.4. Various estimates of annual contribution of CIMMYT to wheat materials in the North (US$ millions)

State/area	Average production (1986-90)[1]	% World production[1]	% North production[1]
World	539448	100	
South	220727	41	
North	318721	98	100
USA	58697	11	18
Australia	14565	3	5
Italy	8391	2	3
New Zealand	249	0	0

State/area	Estimated annual values for CIMMYT wheat material					
	OECD ($500m)[2]	USA ($1800m)[3]	Australia ($75m)[5]	Australia ($122m)[5]	Italy ($122m)[6]	New Zealand ($0.3m)[7]
World						
South						
North	2720	9780	1630	2650	11540	300
USA	501	1801	300	488	2125	55
Australia	124	447	74	121	527	14
Italy	72	257	43	70	304	8
New Zealand	2	8	1	2	9	0

Sources:

1. FAO Agrostat disks, FAO Rome
2. OECD (1982)
3. RAFI extrapolation of data from Dalrymple (1986)
4. Tribe (1991): low estimate
5. RAFI extrapolation of data provided by Tribe (1991): high estimate
6. INTAGRES (1992)
7. Tribe (1991)

would afford Northern donors an even return on their investment. But CIMMYT's US$27 million budget is also for wheat, barley and triticale, which contribute substantially to Northern agriculture. Recently CIMMYT breeders reported that about 30 per cent of requests they receive for farmers' varieties of tropical maize now come from private companies, and that the percentage is rapidly growing.

The US accounted for an average of 54 per cent of the North's dry bean production in the 1986–90 period. CIAT estimates that its materials contribute US$60 million to the US agricultural economy every year. This extrapolates to a Northern gain of about US$111 million from a CIAT 1990 budget of US$28.1 million – a fourfold return.

These figures are crude estimates for only four crops. CIP's contribution in potatoes is likely to be enormous. The International Crops Research Institute for the Semi-Arid Tropics (ICRISAT) virtually established the Australian chickpea industry from 16,000 varieties given to Australian breeders. There are no estimates for barley from CIMMYT, or vegetables from the Asian Vegetable Research And Development Center (AVRDC).

From one perspective, both IARCs and Third World farmers could and should take pride in their contribution to global agriculture. In principle, there is no reason why the North should not benefit. The problem arises when the commercial value flowing North is both not acknowledged and not compensated. The situation is seriously aggravated when Northern governments allow the patenting of material wholly or partially derived from farmers' varieties. As private companies move into Third World seed markets, farmers are finding themselves paying for the end product of their own genius. The North is becoming a huge 'klepto-monopoly', taking freely-given germ-plasm from the South and winning patenting monopolies in the North. Informal innovators, and their NGO supporters, need a policy framework which confronts this situation and which addresses international research systems which are involved in the flow of material and information.

Institutional transformation: the unavoidable responsibility

Any group which seeks validation from the institutional systems is inevitably locked into a relationship with that system. Even if the informal system of farmer and community innovation opted not to seek recognition in the international system, the flow of farmers' genius into the formal system is so great that it would have deep ramification for the informal system, to the extent that a relationship could not be avoided. Within this institutional setting, priority needs to be given to the creation of a strategy to help the researchers transform their system into one which supports the interests of the informal networks of small farmers.

References

Adams, M. E., (1982) *Agricultural Extension in Developing Countries*, Longmans, Burnt Mill.

Alebikiya, M., (1993) 'The Association of Church Development Projects in Northern Ghana', in K. Wellard and J. Copestake, *Non-governmental Organizations and the State in Africa: Rethinking Roles in Sustainable Agricultural Development*, Routledge, London and New York.

Altieri, M.A., (1987) *Agroecology: The Scientific Basis of Alternative Agriculture*, Intermediate Technology Publications, London.

Altieri, M.A. and Merrick, L.C., (1987) 'In-situ conservation of crop genetic resources through maintenance of traditional farming systems', *Economic Botany*, 41 (1): 86–96.

Altieri, M.A. and Yurjevic, A., (1989) 'The Latin American Consortium on Agroecology and Development: A New Institutional Arrangement to Foster Sustainable Agriculture among Resource Poor Farmers', Mimeo, University of California, Berkeley and Centro de Educación y Tecnología, Santiago.

Amanor, K.S., (1990) *Analytic Abstracts in Farmer Participatory Research*, Agricultural Administration Unit Occasional Paper 10, ODI, London.

Amanor, K.S., (1991) 'Managing the fallow: Weeding technology and environmental knowledge in the Krobo district of Ghana', *Agriculture and Human Values*, Vol.8(1–2):5–13.

Amanor, K.S., (1992) 'The New frontier: Ecological management and pioneer settlement in the Asesewa district of Ghana', draft, UNRISD, Geneva.

Arum, G., (1993) 'Kenya Energy and Environment Organizations (1981–90)' in K. Wellard and J. Copestake, *Non-governmental Organizations and the State in Africa: Rethinking Roles in Sustainable Agricultural Development*. Routledge, London and New York.

Ashby, J., (1987) 'The effects of different types of farmer participation on the management of on-farm trials', *Agricultural Administration and (Research and Extension) Network Paper* No 25, ODI, London.

Baker, D.N.D. and Siebert, J., (1984) 'The challenge of developing agriculture in the 400–600 mm rainfall zone within the SADCC countries', *Zimbabwe Agriculture Journal*, 81:205–214.

Barker, D., (1980) 'Appropriate Methodology: Using a traditional African board game in measuring farmers' attitude and environmental images', in D. Brokensha, D.M. Warren and O. Werner (eds) *Indigenous Knowledge Systems and Development*, University Press of America, Lanham.

Bayush, T., (1991) 'Community management of crop genetic resources in the enset-complex farming systems of southern Ethiopia: A case study from Sidamo Region', MSc thesis, NORAGRIC, Agricultural University of Norway.

Bebbington, A. J., (1991) *Farmer Organizations in Ecuador: Contributions to Farmer-First Research and Development*, Gatekeeper Paper No.16, IIED, London.

Bebbington, A. J. and Thiele, G., (1993) *NGOs and The State in Latin America: Rethinking Roles in Sustainable Agricultural Development*, Routledge, London.

179

Berg, T., (1992) 'Indigenous knowledge and plant breeding in Tigray', Ethiopia, *Forum for Development Studies*, Vol.1:13–22.

Berg, T., Bjoernstad, A., Fowler, C., and Kroeppa, T., (1991) *Technology options and the gene struggle*, NORAGRIC Occasional Papers Series C, Development and Environment No. 8., Agricultural University of Norway.

Biggs, S.D., (1988) *Resource-poor farmer participation in research: a synthesis of experiences in nine national research systems.* ISNAR, The Hague.

Biggs, S.D., (1989) 'A Multiple Source of Innovation Model of Agricultural Research and Technology Promotion', *Agricultural Administration (Research & Extension) Network*, Paper No.6, ODI, London.

Biggs, S.D. and Farrington, J., (1991) *Agricultural Research and the Rural Poor: A Review of Social Science Analysis*, IDRC, Ottawa.

Billing, K.J. Chiduza, C., Murphree, M.P. and Reh, I., (1984) Enterprise Patterns and End Uses in the Siabuwa Valley. Sebungwe Regional Study, First Interim Project Report August 1984. Department of Land Management and Centre for Applied Social Studies, University of Zimbabwe, Harare.

Boster, J.S., (1985) 'Selection for perceptual distinctiveness : evidence from Aguaruna cultivars of Manihot esculenta', *Economic Botany*, 39 (3): 310-325.

Boster, J.S., (1986) 'Exchange of varieties and information between Aguaruna manioc cultivators', *American Anthropologist*, 88:428-436.

Botchway, J., (1988) 'The lack of agricultural knowledge networks and institutional frameworks in rural development projects: the case of weija irrigation project.' MSc thesis, Wageningen Agricultural University, Wageningen.

Botswana Technology Center, Food Technology Research Service, (1991) 'Final Report on Morama Beans Processing.' A consultancy report submitted to Thusano Lefatsheng.

Box, L., (1984) 'Adapting agricultural research to small-holder conditions – the case of cassava cultivation in the Dominican Republic.' Paper presented at the Caribbean Studies Association, Conference on Strategies for Progress in the Post Independence Caribbean, St. Kitts.

Box, L., (1986) *Cassava cultivators in the Dominican Republic.* Wageningen Agricultural University, Wageningen.

Box, L., (1986) 'The experimenting farmer: A missing link in agricultural change?' in J. Hinderink and E. Szulc-Dabrowiecka (eds) *Successful Rural Development in Third World Countries*, Netherlands Geographical Studies No 67, Utrecht.

Box, L., (1989) 'Virgilio's theorem: A method for adaptive agricultural research', In R. Chambers, A. Pacey and L. Thrupp, *Farmer First: Farmer Innovation and Agricultural Research*, Intermediate Technology Publications, London.

Box, L. and van Dusseldorp, D., (1992) *Sociologists in agricultural research; Major findings of two research projects in the Dominican Republic and the Philippines*, Wageningen Agricultural University, Wageningen.

Brush, S.B., (1991) 'A farmer-based approach to conserving crop germ-plasm', *Economic Botany*, 45(2):153-165.

Brush, S.B., (1992) 'Farmers' rights and genetic conservation in traditional farming systems', *World Development*, 20(11): 1617-1630.

Brush, S.B., Carney, H.J. and Huaman, Z., (1981) 'Dynamics of Andean potato agriculture' *Economic Botany*, 35(1): 70-88.

Buck, L.E., (1993) 'Development of Participatory Approaches for Promoting Agroforestry: Collaboration between The Mazingira Institute, ICRAF, CARE-Kenya, KEFRI and the Forestry Department' in K. Wellard and J. Copestake (eds), *NGOs and the State in Africa*, Routledge, London.

Bundy, C., (1988) *The Rise and Fall of the South African Peasantry* Heinemann Education, London.

Byerlee, D., Harrington, L., and Winkelmann, D.L., (1982) 'Farming Systems Research: Issues in research strategy and technology design', *American Journal of Agricultural Economics*, December 897:904.

Carroll, T., (1992) *Intermediary NGOs: The Supporting Link in Grassroots Development*, Kumarian Press, West Hartford.

Centre for Community Organization Research and Development, (1991) 'Regaining Control', In M. Ramphele (ed.) *Restoring the Land. Environment and Change in Post-Apartheid South Africa*, Panos Publications, London. pp.65–78.

Chambers, R., (1983) *Rural Development: Putting the last first*, Longman, London.

Chambers, R., (1989) 'Reversals, Institutions and Change.' In: R. Chambers, A. Pacey and L.A. Thrupp (eds) *Farmer First. Farmer Innovation and Agricultural Research*. Intermediate Technology Publications, London, pp.14–24.

Chambers, R, Pacey, A. and Thrupp L.A., (1989) *Farmer First: Farmer Innovations and Agricultural Research*. Intermediate Technology Publications, London.

Chambers, R. and Ghildyal, B.P., (1985) 'Agricultural research for Resource-Poor Farmers: The farmer-first-and-last model', *Agricultural Administration*, 20:1–30.

Charles, R. A. and Wellard, K., (1993) 'Agricultural Activities of Government and NGOs in Siaya District' in K. Wellard and J.G. Copestake (eds), *NGOs and the State in Africa*, Routledge, London.

Checkland, P.B., (1985) 'From optimizing to learning: a development of systems thinking for the 1990s', *Opl. Soc., 36* (9): 757–767.

Clark, J., (1991) *Democratizing development: the role of voluntary organizations.*, Earthscan, London.

Clawson, D.L., (1985) 'Harvest security and intraspecific diversity in tropical agriculture', *Economic Botany*, 39(1): 56-67.

Clayton, E., (1983) *Agriculture, poverty and freedom in developing countries*, London, Macmillan.

Collear, D., (1983) 'Women and coarse grain production in Africa,' *Expert Consultation on Women in Food Production*. ESH :WIFP/83/84, FAO, Rome.

Collinson, M.P., (1972) *Farm Management in Peasant Agriculture: A handbook for Rural Development Planning in Africa*, Praeger, New York.

Collinson, M.P., (1983) 'Technological Potential for Food Production in Eastern and Central Africa', Paper presented at the Conference on Accelerating Agricultural Growth in the Sub-Saharan Africa, Victoria Falls, Zimbabwe.

Conklin, H.C., (1957) *Hanunoo agriculture in the Philippines*, FAO Forestry Development Paper No. 12. FAO, Rome.

Conway, G.R. and Barbier, E.B., (1990) *After the Green Revolution: Sustainable Agriculture for Development*, Earthscan Publications, London.

Cook, O.F., (1925) 'Peru as a center of domestication.' *Journal of Heredity.* 16:33–46; 96–110.

Cordeiro, A., (1991) *Obtenção de variedades de milho para produção de sementes por pequenos agricultores: Estratégias de trabalho da Rede PTA*. Curitiba, AS-PTA Reg. Sul.

Crissman, C.C. (1989) *Seed Potato Systems in The Philippines: A Case Study*, CIP-PCARRD, Lima.

Cromwell, E., (ed.) (1990) 'Seed diffusion mechanisms in small farmer communities: lessons from Asia, Africa and Latin America,' *Agricultural Administration (Research and Extension) Network Paper*, London, ODI.

Cromwell, E., Friss-Hansen, E. and Turner, M., (1992) *The Seed Sector in Developing Countries: A Framework for Performance Analysis*, ODI Working Paper 65, ODI, London.

Dalrymple, D.G., (1986) 'Development and Spread of High Yielding Wheat Varieties in Developing Countries, USAID, Washington, p.96.

Dapper, O., (1668). Cited in A. Jones, (1983). *From Slaves to Palm Kernels: A History of the Galinhas Country (West Africa) 1730–1890*, Wiesbaden, Steiner Verlag.

Dorp, M. van and Utomo, J.S., (1989) 'Consumer Acceptability of Maize.' MARIF/ATA 272, *Internal Technical Report* (mimeo).

Dorp, M. van and Utomo, J.S., (1990) 'Consumer Acceptability of Cassava.' MARIF/ATA 272, *Internal Technical Report* (mimeo).

Dudley, J.W., (1977) '76 generations of selection for oil and protein percentage in maize', *Proc. Int. Conf. on Quantitative Genetics*, 16–21 August 1976, Ames, Iowa State University Press, pp.459–73.

Eisstadt, S.N., (1955) 'Communications systems and social structure: An exploratory comparative study', *Public Opinion Quarterly*, 19:153–167.

Elwell, J., (1991) *Small Farm Development: Understanding and Improving Farming Systems in Humid Tropics of Africa*, Westview Press., Boulder, Co.

Ewel, J.J., (1980) 'Tropical succession: Manifold routes to maturity', *Biotropica*, Vol.12, Supplement, 1980, pp.2–7.

Farrington, J. and Bebbington A., (1993) *Reluctant Farmers: NGOs, the State and Sustainable Agricultural Development*, Routledge, London and New York.

Farrington, J. and Biggs, S.D., (1990) 'NGOs, Agricultural Technology and the Rural Poor', *Food Policy*, 16(1).

Farrington, J. and Martin, A., (1988) *Farmer Participation in Agricultural Research: A Review of Concepts and Recent Practices*, Agricultural Administration Unit Occasional Paper 9, ODI, London.

Ferguson, A.E. and Sprecher, S.L., (1989) 'Designing bean breeding strategies for small-scale farmers in Eastern Africa.' Paper presented at the Symposium on Crop Breeding Criteria and Agricultural Development, American Association for the Advancement of Science Annual Meetings, January 14-19, 1989, San Francisco.

Fowler, C. and Mooney, P.R., (1990) *Shattering*, University of Arizona Press, Tucson.

Fowler, C. and Mooney, P.R., (1990) *The Threatened Gene: Food, Politics and the Loss of Genetic Diversity*, University of Arizona Press, Tucson.

Franco, E. and Schmidt, E., (1985) *Adopción y Difusión de Variedades de Papa en el Departamenta de Cajamarca*, Departmento de Ciencias Sociales, Documento de Trabajo, CIP, Lima.

Frankel, O., (1970) 'Genetic dangers of the green revolution', *World Agriculture* 19: 9-13.

Friss-Hansen, E., (1987) 'Changes in land tenure and land use since villagization and their impact on peasant agricultural production in Tanzania – the case of the southern highlands,' CDR Research report No. 11, Copenhagen, Denmark.

Friis-Hansen, E., (1993) 'Conceptualizing in situ conservation of landraces: the role of IBPGR,' Paper presented at the workshop on the Human, Socio-economic and Cultural Aspects of Plant Genetic Resource Conservation, IBPGR, Rome, 29 April - 1 May.

Goodman, M.M. (1985) 'Exotic Maize Germplasm: Status, Prospects and Remedies', *Iowa State Journal of Research* 59(4), p.501.

Gottstein, G., and Link, G., (eds) (1986) 'Cultural Development Sciences and Technology in Sub-Saharan Africa.' DSE.

Groosman, T., Linnemann, A. and Wierema, H., (1991) 'Seed industry development in a North/South perspective.' Wageningen Agricultural University, Pudoc.

Gubbels, P., (1988) 'Peasant Farmer Self-development: The World Neighbors' Experience in West Africa', *ILEIA Newsletter*, Vol.4, No.3: 11–140.

Guyer, J.I., (1985) 'Review of West African studies', in R. Dixon-Mueller, (ed.), *Women's work in third world agriculture*, ILO, Geneva.

Gyasi, E.A., (1991) 'Communal land tenure and spread of agroforestry in Ghana's Mampong Valley', *Ecology and Farming,* No.2:16–17.

Hanelt, P., (1986) 'Pathways of domestication with regard to crop types (grain legumes, vegetables)', in C. Barigozzi (ed.), *The Origin and Domestication of Cultivated Plants*. Developments in agricultural managed forest ecology, No. 16, Amsterdam: Elsevier, pp.179–200.

Harlan, J.R., (1965) 'The possible role of weed races in the evolution of cultivated plants', *Euphytica*, 14:173-76.

Harlan, J., (1984) 'Gene centers and gene utilization in American agriculture' in Yeatman *et al.* 1984. *Plant Genetic Resources: a Conservation Imperative.* American Association for the Advancement of Science Symposium 87, Washington DC.

Harris, D., (1969) 'Agricultural Systems, Ecosystems and the Origins of Agriculture,' in P.J. Ucko and G.W. Dimbleby (eds) *The Domestication and Exploitation of Plants and Animals.* London.

Hatch, J., (1976) *The Corn Farmers of Motupe: A Study of Traditional Farming Practices in Northern Coastal Peru*, Monograph No. 1, Land Tenure Center, University of Wisconsin, Madison, Wisconsin.

Haugerud, A. and Collinson, M.P., (1990) 'Plants, genes and people: Improving the relevance of plant breeding in Africa,' *Experimental Agriculture,* Vol.26: pp. 341–362.

Haverkort, B., van der Kamp, J., and Waters–Bayer, A., (eds) (1991) *Joining Farmers' Experiments*, Intermediate Technology Publications, London.

Haverkort, B. and Millar, D., (1992) Farmer's experiments and cosmovision, *ILEIA newsletter*, 92(1):26.

Hildebrand, P.E., (1981) 'Combining disciplines in rapid appraisals: The sondeo approach', *Agricultural Administration*, 8:423–432.

Hill, M.H., (1984) 'Where to begin? The place of the hunter founder in Mende histories', *Anthropos* 79, pp.653–656.

Hodgkin, T. and Ramanatha Rao, V., (1992) 'IBPGR and the conservation of landraces', Discussion paper, IBPGR, Rome.

Horton, D. and Prain, G., (1989) *Beyond FSR: New Challenges for Social Scientists in Agricultural R and D*, mimeo from the International Potato Centre (CIP).

House, L.R., (1987) 'Sorghum and food security in Southern Africa: present and future research priorities of technical scientists', in M. Rukuni and C.K. Eicher (eds), *Food Security for Southern Africa*, UZ/MSU Food Security Project, Department of Agricultural Economics and Extension, University of Zimbabwe, Harare.

Hyman, H., (1963) 'Mass media and political socialization: the roles of patterns of communication', in L.W. Pye (ed.) *Communications and political development*, Princeton, Princeton University Press, pp.128–151.

ICRISAT, (1990) *Annual Report*, Patancheru, Andhra Pradesh, India.

Ingram, B.G. and Williams, J.T., (1984) '*In situ* conservation of wild relatives of crops',in J.H.W. Holden and J.T. Williams (eds) *Crop Genetic Resources: Conservation and Evaluation*, George Allen and Unwin, London.

INTAGRES (1992) (D. Bagnara), 'Developing Countries Contribute Germplasm to Italian Plant Bredding Programs through the CGIAR Centres', *Diversity* 8:1.

International Crops Research Institute for the Semi-Arid Tropics, (1980/83), *Annual Report*, Ouagadougau, Haute-Volta, Programme Cooperatif. ICRISAT/Haute-Volta.

IPGRI, (1993) *Diversity for Development: The Strategy of the International Plant Genetic Resources Institute*, IPGRI, Rome.

Janzen, D.H., (1980) *Ecology of Plants in the Tropics*, Edward Arnold, London.

Jintrawet, A., Smutkupt, S., Wongasamun, C., Kersuk, V., (1985) 'Extension activities for peanuts after rice in Ban Sum Jan, Northeast Thailand: A case study in farmer-to-farmer extension methodology', draft, Khon Khaen University, Thailand.

Kaluli, J., (1993) 'NGO involvement in agricultural activities in Machakos District' in K. Wellard and J. Copestake (eds). *NGOs and the State in Africa*, Routledge, London.

Keystone Center, (1991) 'Global initiative for the security and sustainable use of plant genetic resources' final consensus report of the Keystone International Dialogue Series on Plant Genetic Resources, Third plenary session, 31 May - 4 June, 1991, Oslo, Norway, Keystone Center, CO.

Kiambi, K., and Opole, M., (1992) 'Promoting traditional trees and food crops in Kenya.' in D. Cooper, R. Vellvée and H. Hobbelink (eds) *Growing Diversity*, Intermediate Technology Publications, London.

King, S.R., (1988) *Economic Botany of the Andean Tuber Crop Complex: Lepidium meyenii, Oxalis tuberosa, Tropaeolum tuberosum and Ullucus tuberosus*. PhD Dissertation. The City University of New York.

Kloppenberg, J. and Kleiman, D., (1987) 'The plant germ-plasm controversy. Analyzing empirically the distribution of the world's plant genetic resources', *BioScience*, 37(3): 190-198.

Kolbilla, D., and Wellard, K. (1993) 'Langbensi Agricultural Experimental Station: experiences of research', in K. Wellard and J. Copestake (eds) *Non-*

governmental organizations and the State in Africa. Routledge, London and New York.

Korten, D.C. and Klauss, R., (1984) *People-centred Development: Contributions Toward Theory and Planning Frameworks*. Kumarian Press, West Hartford.

Korten, D., (1990) *Getting to the twenty-first century: voluntary development action and the global agenda*. Kumarian Press, Connecticut.

Kortlandt, A., (1984) Vegetation research and the 'bulldozer' herbivores of tropical Africa. In A.C. Chadwick and S.L. Sutton, (eds) *Tropical rain forest*, Special Publication of the Leeds Philosophical and Literary Society, pp. 205–226.

Kuhn, T., (1970) *The structure of scientific revolutions*, Chicago University Press, Chicago.

Latour, B., (1983) 'Give me a laboratory and I will raise the world', in K.D. Knorr-Cetina and M. Mulkay (eds) *Science observed: perspectives on the social study of science*, Sage, London.

Lawton, H.W. and Wilke, P.J., (1979) 'Ancient agricultural systems in dry regions' in A.E. Hall, G.H. Cannell and H.W. Lawton (eds) *Agriculture in Semi-arid Environments*, Springer, New York, pp. 1–44.

Lipton, M.,(1968) 'The theory of the optimizing peasant', *Journal of Development Studies*, 4 : 327-351.

Lipton, M. and Longhurst, R., (1989) 'Modern varieties, international research and the poor,' CGIAR Study Paper No. 2. The World Bank, Washington.

Long, N., (1977) *An introduction to the sociology of human development*, Tavistock, London.

Long, N., (1990) 'From paradigm lost to paradigm regained? The case for an actor-oriented sociology of development', *European Review of Latin American and Caribbean Studies*, Vol.49 (December).

Louette, D., (1992) Estudio del manejo campesino de variedades locales de maíz cultivada en la Reserva de la Biosfera Sierra de Manantlan R.S.B.M, estados de Jalisco y Colima, México. Que conservamos in situ? Trabajos realizados entre noviembre de 1989 y febrero de 1992. Reserva de la Biosfera Sierra de Manantlan, Laboratorio Natural Las Joyas, Universidad de Guadalajara, Mexico.

McCorkle, C.M., Brandsletter, R.H., and McClure, G., (1988) *A case study on farmer innovation and communication in Niger*, Communication for Technology Transfer in Africa, Academy of Educational Development, Washington.

Machado, A.T., Magnavaca, R., (1991) *Estresse Ambiental: O milho em perspectiva*. Rio de Janeiro: AS-PTA, (Cardernos de T.A.).

Mann, M., (1983) *Macmillan Student Encyclopedia of Sociology*, Macmillan Press, London.

MARIF/ATA, (1990) 'Technology Developed for Farmers,' Research Proposal 'Farmers' Criteria in the Selection of Varieties of Palawija Crops. An Assessment Study.' (mimeo).

Mateo, N. and Tapia, M., (1990) 'High mountain environment and farming systems in the Andean region of Latin America.' In *Mountain Agriculture and Crop Genetic Resources*. International Workshop IDRC/ICIMOD, 1987. Kathmandu, Nepal. Oxford Publications, New Delhi.

Matlon, P.J, Cantrell, R., King, D., and Benoit-Cattin, H., (eds) (1984) *Coming Full Circle: Farmers' participation in the development of technology*, IDRC, Ottowa.

Matlon, P.J. and Spencer, D.S., (1984) 'Increasing food production in sub Saharan Africa: environmental problems and inadequate technical solutions', *American Journal of Agricultural Economics*, 66(5), pp.671–676.

Maurya, D.H., Bottrall, A., and Farrington, J., (1988) 'Improved livelihoods, genetic diversity and farmer participation: a strategy for rice breeding in rainfed areas in India', *Experimental Agriculture*, 24(3) Farming Systems series – 14:311-320.

'Mbewe, M.N., (1992b) 'The distribution, utilization and conservation of some under-utilized plant genetic resources in Botswana'. Paper presented at National Workshop on Plant Genetic Resources, Gaborone Sun, Gaborone, Botswana, 23 – 26 June, 1992.

'Mbewe, M.N., (1992a) The domestication of morama beans (*Tylosema esculentum*, Fabaceae), Paper presented at the International Conference on the development of New Crops, Sheraton Hotel, Jerusalem, Israel, 8 – 12 March, 1992.

Mellor, J.W., (1966) *The Economics of Agricultural Development*, University of Cornell Press, Ithaca, New York.

Merrick, L.C., (1990) 'Crop genetic diversity and its conservation in traditional agroecosystems' in M. Altieri and S. Hecht (eds) *Agroecology and Small Farm Development*, CRC Press, Boston.

Millar, D., (1992) 'Understanding Rural Peoples' Knowledge and its Implications for Intervention: From the roots to the branches', Case studies from Northern Ghana, MSc thesis, Wageningen Agricultural University, Wageningen.

Moss, H. and Taylor, F.W., (1981) *The potential for commercial utlization of veld products*. Ministry of Commerce and Industry of Botswana. Government Printer, Gaborone.

Mushita, T.A., (1991) *Development of Techniques for Sustainable Agriculture Production in Semi-Arid Areas of Zimbabwe*, ENDA-Zimbabwe, Harare.

Mushita, T.A., (1992) 'Farmers' knowledge in the selection and conservation of indigenous crops in Zimbabwe.' Paper presented at the seminar on Local Knowledge and Agricultural Research, organized by Wageningen Agricultural University and ENDA-Zimbabwe, held at Brondesbury Park, Zimbabwe, 28 September–2 October 1992.

Nabham, G.P., (1985) 'Native crop diversity in Aridoamerica: Conservation of regional gene pools', *Economic Botany*, 39 (4):387-99.

Nair, P.K.R. (ed.) (1986) *Agroforestry systems in the tropics*, Kluwer, Dordrecht.

Nazarea-Sandoval, V.D., (1991a) 'If The Shoe Fits: Local criteria for varietal selection and certain implications for agricultural research and development.' Paper presented for the UPWARD Network for the International Potato Centre, The Philippines.

Ndiweni, M., (1993) 'The Organization of Rural Associations for Progress and Grassroots Development' in K. Wellard and J. Copestake (eds), *NGOs and the State in Africa*, Routledge, London.

Norgaard, R.B., (1984) 'Traditional Agricultural Knowledge: past performance, future prospects and institutional implication', *American Journal of Agricultural Economics*, 66:874–878.

Norman., D. (1974) 'Rationalizing mixed cropping under indigenous conditions: 'The example of Northern Nigeria', *Journal of Development Studies*, 11(1):3–21.

NRC, (1989) *Lost crops of the Incas*, NRC, New York.

NRC, (1992) 'Conserving Biodiversity. A research agenda for development agencies.' Report of a Panel of the Board on Science and Technology for International Development U.S. National Research Council, National Academy Press, Washington, DC.

Nye, P.H. and Greenland, D.J., (1960) *The soil under shifting agriculture,* Commonwealth Bureau of Soils Technical Communications, No.52, Commonwealth Agricultural Bureau, London.

Nye, P.H. and Stephens, D., (1962) 'Soil Fertility' in J.B. Wills (ed.) *Agriculture and Land Use in Ghana,* Oxford University Press, London, p.127–143.

Nyerges, A.E., (1987) 'The development potential of the Guinea savanna: Social and ecological constraints in the West African 'middle belt" in P.D. Little, M.M. Horowitz and A.E. Nyerges (eds) *Lands at Risk in the Third World: local-level perspectives,* Westview Press, Boulder and London.

Nyerges, A.E., (1989) 'Coppice swidden fallows in tropical forest: Biological, technological, and sociocultural determinants of secondary forest succession', *Human Ecology,* Vol.17 (4), 1989, pp.379–400.

Obilana, T.A., (1988) Progress in Sorghum breeding in the SADCC Region (1984/85 – 1987/88), and plans for 1988/89, in *Proceedings of the fifth Annual Workshop on Sorghum and Millets for Southern Africa,* 21–23 Sept 1988, Maseru, Lesotho, pp.213–248.

Okeyo, A.P., Opole, M., (1988) 'The indigenous vegetables project.' A report for KENGO, December.

Oldfield, M. and Alcorn, J., (1987) 'Conservation of traditional agroecosystems', *BioScience,* 37:199-208.

Oomkes, F.R., (1986) *Communicatieleer,* Meppel, Bonn.

Opole, M., (1989) 'Sustainable development and education. – The role of women in energy and food production.' In Daysh, M. Carley, E. Ekehorn, K. Phillips and R. Walter (eds), The Ninth Commonwealth Conference on Human Ecology, CHEC, UK

Opole, M., (1992) 'Indigenous knowledge and appropriate technology.' A report for Kenya Forestry Masterplan, Forestry Department, Government of Kenya, March,

Opole, M., (1991) 'Women's indigenous knowledge base in the translation of nutritional and medicinal values of edible local plants in Western Kenya,' Manuscript.

Opole, M., Chweya, J. and Imungi, J., (1991) 'Indigenous vegetables of Kenya:- A two year field and laboratory experience of research.' KENGO, Nairobi.

Owusu, D.Y., (1990) 'Experiences with agroforestry', *ILEIA Newsletter,* 6(2):810.

Paterniatri, E, and Viegas, G.P., (1987) *Melhoramento e produção de milho.* Campinas, Fundação Cargill.

Ploeg, J.D. van der, (1990) *Labor Markets, and Agricultural Production.* Westview Press, Boulder.

Plucknett, D.L., Smith, N.J.H., Williamns, J.T.Y and Anishetty, N.M (1987) *Gene Banks and the Worlds' Food,* Princeton University Press, New Jersey.

Pointing, C., (1991) *Green history of the world,* London, Penguin.

Popper, K.R., (1963) in D.L. Sills (ed) *International encyclopedia of the social sciences,* Vol. 12, Collier-Macmillan Press, London, pp.159–164.

Prain, G., (1992) 'Cultivators, contexts and cultural knowledge: Interdisciplinary experiences in the collection of Ipomea Batatas Germ-plasm in Latin America', paper prepared for the five day workshop entitled – Local Knowledge and Global Resources: Involving Users in Germ-plasm Conservation and Evaluation, Alaminos, Pangasinan.

Prain, G. and Scheidegger, U., (1988) *User-Friendly Seed Programs*, Report of the Third Social Science Planning Conference, held at the CIP, Lima, Peru, September, 7–10, 1987.

Prain, G. and Uribe, F., (1990) 'The Trials and Errors of Potato Seed Distribution,' in Cromwell, E. (ed.) *Seed Diffusion Mechanisms in Small Farmer Communities: Lessons from Asia, Africa and Latin America*, Agricultural Administration (Research and Extension) Network, Network Paper No.21, ODI, London.

Prain, G., Uribe, F., and Scheideger, U., (1992) 'The Friendly Potato: Farmers Selection of Potato Varieties for All Occasions,' in J. Moock and R. Rhoades, (eds) *Diversity, Farmer Knowledge and Sustainability*, Cornell University Press, New York.

PRATEC, (1991) 'Andean Agriculture and Peasant Knowledge: Revitalising Andean Knowledge in Peru', in B. Haverkort, J. van der Kamp, and A. Waters-Bayer (eds) *Joining Farmers' Experiments*, Intermediate Technology Publications, London, pp.93–112.

Prescott-Allen, R. and Prescott-Allen, C., (1982) 'The case for *in-situ* conservation of crop genetic resources', *Nature and Resources*, Vol.23:15-20.

Pye, L.W., (1963) *Communications and political development*, Princeton.

Reijntjes, C., Haverkort, B., and Waters-Bayer, A. (1992) *Farming For the Future: An introduction to low-external-input and sustainable agriculture*, Macmillan, London.

Reimann, H., (1974) *Komminikations-systeme: Umriss einer Sociologie der Vermittlungs-und Mitteilungsprozesse*, Tubingen, JCB Mohr (Paul Siebeck).

Rhoades, R.E., (1982) 'Understanding small farmers: sociocultural perspectives on experimental farm trials.' Social Science Department training document 1982–3, CIP, Lima.

Rhoades, R.E., and Bebbington, A., (1991) 'Farmers as experimenters' in B. Haverkort, J. van der Kamp and A. Waters-Bayer, (eds) *Joining Farmers' Experiments*, Intermediate Technology Publications, London.

Rhoades, R.E. and Booth, R.H., (1982) 'Farmer-back-to-farmer: A model for generating acceptable agricultural technology', *Agricultural Administration*, 11:127–137.

Rhoades, R. E., Horton, D. H., Booth, R. H., (1987) Anthropologist, biological scientist and economist: the three musketeers or three stooges of farming systems research?', in J.R. Jones and B.J. Wallace (eds) *Social science and farming systems research: methodological perspectives on agricultural development*, Westview Press, London, pp.21–42.

Richards, P., (1985) *Indigenous agricultural revolution. Ecology and food production in West Africa*, Hutchinson, London.

Richards, P., (1986) *Coping with Hunger: Hazard and Experiment in an African Rice Farming System*, Allen and Unwin, London.

Richards, P., (1987) 'Experimenting Farmers and Agricultural research', Draft, Dept. of Anthropology, University College London.

Richards, P., (1990) 'Local strategies for coping with hunger: northern Nigeria and central Sierra Leone compared,' *African Affairs*, 89, pp.265–275.

Richards, P., (1991) 'Mende names for rice: cultural analysis of an agricultural knowledge system,' Conference proceedings, *Agricultural Knowledge systems and the role of extension Stuttgart*. Institut für Agrar- und soziolokonomie, Universität Hohenheim.

Richards, P., (1993) 'Natural symbols and natural history: chimpanzees, elephants, and experiments in Mende Thought', in K. Milton (ed.) *Environmentalism: the view from anthropology*, ASA Monograph 32, Routledge (in press), London.

Rocheleau, D., Wachira, K., Malaret, L., Wanjohi, B.M., (1989) 'Local knowledge for agroforestry and native plants' Manuscript.

Röling, N., (1990) 'The agricultural research technology transfer interface: A Knowledge System Perspective,' in D. Kaimowitz (ed.) *Making the Link: Agricultural Research and Technology Transfer in Developing Countries*, Westview Press, Boulder.

Rozas, J.W., (1985) *El sistema agricola andino de la CC de Amnaru*. Tesis Fac. de Antropologia. Universidad del Cusco, Peru.

Ruthens, T. *et al.*, (1989/90) Various internal reports on germ-plasm collection trips in East and Central Java. MARIF/ATA 272 (mimeo).

Sandoval, V.N., (1990) *Memory Bank of the Indigenous Technology of Local Farmers Associated with Traditional Crop Varieties: Focus on Sweet Potato*, UPWARD Inaugural Planning Workshop, Hyatt, Baguio.

SAREC, (1992) 'Future Food Security and Plant Genetic Resources.' Report on a Consultation on a Global System for the Security and Sustainable Use of Plant Genetic Resources. SAREC Conference Report 1992:2, Stockholm, Sweden.

Scherr, S. and Muller, E., (1991) 'Evaluating technology performance in agroforestry', *ILEIA Newletter*, 1991 (1): 34–35.

Shiva, V. and Dankelman, I., (1992) 'Women and biological diversity: lessons from the Indian Himalayas,' in D.Cooper, R.Vellve, and H.Hobbelink (eds) *Growing Diversity: Genetic resources and local food security*, Intermediate Technology Publications, London, pp.44–52.

Shone, A.K., (1979) 'Notes on the morula'. Bulletin 58, Department of Forestry, Pretoria, Republic of South Africa.

Simaraks, S., Khammeang, T. and Uriyapongson, S., (1986) 'Farmer to farmer workshop on samll dairy cow raising in three villages, northeast Thailand', draft, Khon Kaen University, Thailand.

Simmonds, N.W., (1979) *Principles of Crop Improvement*. Longman, London.

Smeathman, H., (1783), Appendix to C.B. Wadstrom *An essay on colonization*, London.

Sperling, L., Loevinsohn, M.E. and Ntambovura, B. (forthcoming) 'Rethinking the farmer's role in plant breeding: local bean experts and on-station selection in Rwanda', *Experimental Agriculture*.

Tan, J.G., (1986) 'A participatory approach in developing an appropriate farming system in eight irrigated lowland villages', in C.B. Flora and M. Tomecek (eds), *Selected Proceedings of Kansas State University's Farming System Research Symposium*, Kansas.

Taylor, D., (1988) 'Agricultural Practices in Eastern Maputaland.' *Development Southern Africa.* Vol. 5, No. 4, pp 465–481.

Taylor, D., (1991a) 'Towards Sustainable Agriculture: A people Orientated Approach,' In *Agroforestry: Emphasis on Southern Africa.* Environmental Forum Report, Foundation for Research Development, Pretoria, pp.178–187.

Taylor, D., (1991b) 'Sustainable Agriculture and Technology Transfer: An Agroecological Question for a People Orientated Solution,' Proceedings of the Africa International Conference on Environment, Technology and Sustainable Development, Mozambique, Maputo.

Tessema, T., (1986) 'Improvement of indigenous durum wheat landrace in Ethiopia', In J.M.M. Engels (ed.) *Proceedings of the International Symposium on the Conservation and Utilization of Ethiopian Germ-plasm,* PGRC/E Addis Ababa, pp.232–238.

Thayer, J.S., (1983) 'Nature, culture and the supernatural among the Susu' *American Ethnologist,* Vol.10, No. 1, pp. 116–132.

Theodorson, G.A. and Theodorson, A.G., (1969) *Modern Dictionary of Sociology,* Thomas Y. Cromwell, New York.

Thrupp, L.A., (1991) 'Dynamics of indigenous knowledge for sustainable agriculture: from marginalization to participation and mobilization of rural people', Conference paper presented at *Varieties of sustainability: ethics, ecological soundness and economic equity,* Pacific Grove, California, May 10–12, 1991.

Tribe, D. (1991) *Doing Well by Doing Good - Agricultural Research: Feeding and Greening the World,* Pluto Press, Victoria.

Tuffuor, K., (1992) 'Role of Forestry in Biological Diversity Conservation in Ghana', Paper presented at Biodiversity Workshop at Department of Botany, University of Ghana, Legon, January 8 1992.

Tylor, E.B., (1871) *Primitive culture,* Vol. 1, John Murray, London.

UNEP, (1992) *Convention on Biological Diversity,* UNEP, Nairobi.

Vaughan, D.A., and Chang, T.T., (1992) 'In situ conservation of rice genetic resources', *Economic Botany,* 46(4) 368-383.

van Dusseldorp, D.B.W.M., (1992) *Projects for rural development in the Third World: Preparation and implementation,* Wageningen Agricultural University, Wageningen.

Vavilov, N.I., (1951) 'Plant breeding as a science; Introductory essay to a selection of Vavilov' writings,' In *The Origin, Variation, Immunity and Breeding of Cultivated Plants, Chronica Botanica,* Vol.1/6 pp.1–12.

Viegas, G.P., (1989) *Melhoramento de milho para condições adversas,* (Fundação Cargill), Campinas.

Voss, J., (1992) 'Conserving and increasing on-farm genetic diversity: farmer management of varietal bean mixtures in Central Africa' in J.L. Moock and R.E. Rhoades (eds), *Diversity, farmer knowledge, and sustainability,* Cornell University Press, Ithaca.

Warren, D.M., (1991) *Using indigenous knowledge in agricultural development,* World Bank Discussion Paper No. 127.

Watts, M., and Carney, J., (1990) 'Manufacturing dissent: work, gender and the politics of meaning in a peasant society,' *Africa,* 60 (2), pp. 207–41.

Weber, M., (1980) *Wirtschaft und Gesellschaft: Grundriss der versehenden sociologie*, Tubingen, JCB Mohr (Paul Siebeck).

Wehmeyer, A.S., (1980) 'Some Botswana veld foods which could possibly be used on a wider scale.' National Food Research Institute, Division of Food Chemistry, Pretoria, Republic of South Africa.

Weija Irrigation Project, (1976) Feasibility report, Tahal consultants, Accra, 1976.

Weiskel, T.C., (1989) 'The ecological lessons of the past: an anthropology of environmental decline', *The Ecologist*, 19 (3):98–103.

Wellard, K., and Copestake, J., (eds) (1993 *NGOs and the State in Africa. Rethinking Roles in Sustainable Agricultural Development*, Routledge: London.

Wener, M., Opole, M., Mwangi, M., Wainaina, J., and Ogola, B., (1992) 'Socio-economic and participatory aspects.' Kenya Forestry Masterplan Forest Department, GOK, June.

Wilairat, A., (1985) 'Sesame before rice: A potential cropping system for rainfed farmers in northeast Thailand', Draft, Khon Kaen University, Thailand.

Wilde J.C., (1967) *Experiences with Agricultural Development in Tropical Africa*, John Hopkins University Press, Baltimore.

Wilkes, H.G., (1977) 'Hybridization of maize and teosinte, in Mexico and Guatemala and the improvement of maize', *Economic Botany*, 31(3):254-93.

Wilkes, H.G., (1983) 'Current status of crop plant germ-plasm', *Critical Review of Plant Science*, 1:133-81.

Wood, D., and Lenné, J., (1993) 'Dynamic management of domesticated bio-diversity by farming communities.' Paper presented at the UNEP Expert Conference on Biodiversity, Trondheim, Norway, 24-28 May 1993.

Worede, M., (1974) 'Genetic improvement of quality and agronomic character-istics of durum wheat for Ethiopia', PhD dissertation, University of Nebraska.

Worede, M., (1986a) Conservation and utilization of annual oilseed genetic resources in Ethiopia, *Proceedings of Third Oil Crop Network Workshop (IAR/IDRC)*, Addis Ababa, 6–10 October 1986.

Worede, M., (1986b) An Ethiopian perspective on the conservation and utilization of crop genetic resources, proceedings of *First International Symposium on Conservation and Utilization of Ethiopian Germ-plasm*, Addis Ababa, 13–16 October 1986.

WRI, IUCN and UNEP (1992) *Global biodiversity strategy. Guidelines for actions to save, study, and use earth's biotic wealth sustainably and equitably*. WRI, IUCN and UNEP.

Glossary

Accession	An individual sample of seeds or plants entered into a germ-plasm collection in a genebank (e.g. one accession can consist of a number of seeds or one variety (line or population).
Adaptation	The process by which individuals, populations, or species change in form or function in such a way to better survive under given environmental conditions.
Agro-ecosystem	An ecological system modified by people to produce food and other products for human use.
Allele	One of a pair or series of forms of a gene which are alternative in inheritance because they are situated at the same position occupied by a gene in chromosomes.
Character (in plant breeding)	The sum of the total of the external conditions which affect the growth and development of an organism.
Ecosystem	The dynamic complex of micro-organism, plant, and animal (including human) communities and their non-living environment, interacting as a functional unit.
Ecozone	Area of transition between two ecological systems.
Environment (in plant breeding)	The sum of the total of the external conditions which affect growth and development of an organism.
Ex situ conservation	The conservation of components of biological diversity outside their natural habitats; in the case of plant genetic resources, this may be in genebanks, or in other live collections. In this book, it is especially used to indicate conservation in

193

genebanks.

Farmer participatory research	A framework for research which involved farmers in various processes of research in order to identify, design, test and evaluate new technologies which are appropriate to the needs of small-scale farmers.
Farming systems research (FSR)	Adaptive research in which multi-disciplinary teams are used to assess the technological needs of an area, and to target technology development to specific groups or specific areas. The testing of technology by researchers is conducted on-farm rather than in experimental stations.
Gene	The unit of inheritance. Genes are located at fixed loci (single locus) and can exist in a series of alternative forms called alleles.
Genetic diversity	The genetic variability within a species.
Genetic resources	The genetic material of actual or potential value (germ-plasm in strict plant breeding sense). In a broad sense, the germ-plasm plus information and local knowledge on the germ-plasm (germ-plasm in a socio-economic and cultural perspective).
Genotype	The entire genetic constitution of an organism.
Homozygous	Having alike alleles at corresponding loci on homologous chromosomes.
Hybrid variety	A variety produced in a crossing of two or a complex crossing of various parent lines, resulting in a combination of characters which cannot be maintained by farmers in succeeding generations, and therefore must be purchased each year.
Hybrid	The product of a cross between genetically unllike parents (in strict plant breeding sense).
Indigenous knowledge (IK)	See local knowledge.
In situ conservation	The conservation of ecosystems and natural habitats and the maintenance and recovery of viable populations of species in their natural surroundings. In the case of domesticated or cultivated species, such as

194

	crops, conservation in the surroundings where those populations have developed their distinctive properties. In this book it is used to designate conservation of plant genetic resources on the farm and by communities, where they have been developed, bred and maintained.
Introgression	The flow and exchange of genes between plant populations within (e.g. between landraces and modern varieties) and between species (e.g. between crop species and related weedy species).
Landrace	A cultivated variety, developed and maintained by farmers, either as a line, mixture of lines or population. Landraces behave as varieties which are continuously developing under selective human interference through an active and constant process of adaptation to local production stress and product preferences specific to different groups within farming communities, relating to social differentiation, gender and ethnicity.
Line (in plant breeding)	A community of individuals which has been produced by a process of continued self-fertilization, accompanied by selection. Pure lines are homozygous at all loci. Varieties are mostly lines in the case of self-pollinating crops. In-bred lines are used for the production of hybrid varieties.
Local crop development	The continuous and dynamic process of maintenance, development and adaptation of germ plasm to the environment, and local agro-ecological production conditions, and specific household needs related to social differentiation, gender and ethnicity.
Local knowledge	The knowledge of a people of a particular area based on their interactions and experiences within that area, their traditions, and the incorporation of knowledge emanating from elsewhere into their production and economic systems.
Mass selection	A simple plant breeding technique whereby individual plants are selected and the next generation propagated from the aggregate of

	their seeds.
Matrix ranking	Using matrices and physical objects such as seeds (to construct the matrices) to develop a scoring system to evaluate a complex series of problems).
Modern variety	The product of formal, institutional and scientific plant breeding applying modern techniques of selection and technology and resulting in mostly homogeneous varieties (lines for self-pollinating crops, hybrid or open pollinated varieties of cross-pollinating crops, and clones for vegetatively reproduced crops).
Participatory rural appraisal	A research approach in which techniques are used to help rural people analyse problems and develop their own solutions. Research methods are similar to RRA but the main emphasis is on the researcher facilitating communities to analyse their situation, rather than on communities facilitating researchers to make their own analysis.
Participatory technology development	See farmer participatory research.
Passport data	Information registered at the time of collection of germ-plasm, together with identifying names and numbers, stored in an information system of a genebank.
Phenotype	Appearance of individual as contrasted with its genotype. Used to designate a group of individuals with similar appearance but not necessarily identical genotypes.
Polygenic trait	A character achieved by the effect of genes which cannot be identified individually, but which, through supplementary effects, can have important effects on the total variability of the trait.
Population (in plant breeding)	A community of individuals which share a common gene pool (e.g. a crossing made between individual genotypes).
Qualitative trait	A character in which variation is discontinuous (e.g. major gene resistance, certain morphological characters like flower

colour).

Quantitative trait	A character in which variation is continuous so that classification into distinct categories is difficult (e.g. yield, adaptability, drought, cold tolerance).
Rural people's knowledge (RPK)	See local knowledge.
Seasonality ranking	Using matrices, diagrams, and physical objects to document systematically the rhythm of activities throughout the year and seasonal change, including rainfall, labour, diet, sickness, prices etc.
Selection (in plant breeding)	Discrimination among individuals in the number of offspring contributed to the next generation.
Time lines	A research method which focuses on charting events from the past, on facilitating the systemization of remembrance of events and chronologies of events.
Transects (sweeping or combing)	Walking through an area with local informants to define systematically particular features and problems, such as patterns of land use, technologies, environmental factors, and rights in land, which can be used for diagramming and mapping.
Transfer of technology approach (TOT)	A top-down framework for agricultrual research, which fails to recognize a role for farmers in research and is characterized by poor feedback from farm to research station. Standardized technology is developed in international research centres, adapted in national research centres and disseminated to farmers through extension services, without farmers playing a role in defining problems, nor in designing and evaluating technology.
Trend diagrams	The use of diagrams to help people analyse changes which have happened in the recent past in such areas as ecology, cropping systems, resource utilization, etc.

197

This Glossary is based on the following documents, which can also serve as an introduction to the fields covered in the present book:

Allard, R. W. (1960), *Principles of Plant Breeding*. John Wiley and Sons, New York.

Ford-Lloyd, B. and Jackson, M. (1986), *Plant Genetic Resources: an introduction to their conservation and use*. Edward Arnold, London.

Reijntjes, C., Haverkkort, B. and Waters-Bayer, A., (1992) *Farming for the future: an introduction to low-external-input and sustainable agriculture*. Macmillan Press, London.

Index

199

hunting, 28, 29, 32
hybridization, 3, 69, 83
Hyman, H., 20

ICRISAT, 88, 92, 178
ideology, 19, 54, 155, 163, 164
IFAP, 11
in-breeding, 74
income, 29, 31, 70, 92, 129
India, 11, 100, 128–35
　Myrada PIDOW project, 11, 100, 129–35
　Vegetable Project, 158–9, 163–4
Indonesia, 99, 105, 119–27
　MARIF, 119
industrial crops, 65
Industrious Mothers, 109
industry, 65
informal sector, 88, 92, 178
information, exchange of, 9, 21, 48–9, 86,
　115, 139, 140, 142, 158
infrastructure, 41, 88, 146
Ingram, B.G., 68
INIAA, 115
innovation, farmer, xiii, 4, 6, 7, 9, 18, 19, 36,
　40–1, 45, 55, 57, 100, 156, 174, 178,
　see also experimentation
inputs, 2, 4, 5, 7, 12, 18, 28, 32, 36, 64–6
　passim, 140–1, 148, 157, 162, 164
insecurity, food, 85
institutions, 1, 4, 9, 11, 17–19 *passim*, 24,
　26, 66, 100, 128, 135, 150–1, 178
INTAGRES, 176
intellectual property rights, 4, 13, 141, 173–5
interbreeding, 81
interest rates, 76, 129
International Coalition for Development
Action Seeds Campaign, 12
interviews, 36, 90, 105, 120–1
introgression, 3, 69, 81, 93
IPGRI, 2
irrigation, 2, 27–34, 64, 109, 148
　Weija Project, 17–18, 27–34
ISAR, 10
Islam, 54–5
Italy, 175, 176

Janzen, D.H., 41
Jintrawet, A., 7
Jones, A., 52

Kaluli, J., 139
KENGO, 139, 158
Kenya, 139, 159–64
Keystone Center, 2, 4, 66, 68, 95
Khon Kaen University project, 7
King, S.R., 111

Klaus, R., 34
Kleinman, D., 1
Kloppenberg, J., 1
knowledge, local, xii, 7, 13, 17–57, 61, 65,
　69, 72–84, 95, 99–100, 102–10, 126,
　130–42, 146, 148–9, 151, 155–6, 158–64,
　172–8
　AKIS, 50
　documenting/disseminating, 140, 141
　NGOs and, 137–42
　RPK, 49–50
Kolbilla, D., 10, 140
Korten, D.C., 13, 34
Kortlandt, A., 52
Kuhn, T., 23

labour, 18, 23, 29–32, 42, 48, 51, 52, 56,
　57, 88, 91, 92
　constraints, 18, 29–32, 42, 52
　gender division of, 18, 30–1, 34, 51, 125
　family, 30–2, *passim*
land
　'commons', 130
　dispossession of, 85, 146
　policies, 18, 37–43, 52, 85, 130
　scarcity, 36–7, 41
　use, 28, 85, 112, 114, 147
landless, 129
landraces, 1–4 *passim*, 9, 51, 61–95, 115,
　122, 126, 148, 149, 173 *see also* varieties,
　wild
　elite, 79, 82, 83
Langbensi Agricultural Station, Ghana, 140
Latin America, xii, xiii, 9, 69, 99, 104–8,
　111–18, 139–41 *passim*
　CLADES, 7, 139
Latour, B., 44
Lawton, H.W., 68
legislation, 4, 12, 13, 85, 155, 171
legumes, 65, 66, 113
Lenné, J., 2, 4
Leucaena spp., 39, 42
linkages, xiii–xiv, 9, 99–151
　farmers/researchers, xii, 4, 9–11, 13, 17,
　20–6, 62, 70–1, 93, 99–110, 130–42,
　160–2
Lipton, M., 2
livestock, 74, 124, 137, 147 *see also* cattle-
　rearing
loans, 33–4, 129
　seed, 76
Long, N., 32, 44, 50
Longhurst, R., 2
Longley, Catherine, 17, 19, 51–7
Louette, D., 3
Lynam, J., 90

maintenance, 3, 67, 69, 82
maize, 2, 3, 12, 28, 31, 37, 39, 41, 73, 79,
  80, 85, 86, 112, 117, 119, 121–3 *passim*,
  149, 156, 160, 165–71, 175, 176, 178
  CIMMYT, 175–6
Mali, 10
malnutrition, 85
Mann, M., 26
manure, 33, 132, 149, 161–2
mapping, 11, 131, 134–5
  mental, 105
marginalization, 1, 9, 17, 85, 100–1, 146,
  155, 165
marketing, 18, 30–2 *passim*, 66, 84, 99
markets, seed, 156
Mascarenhas, James, 11, 99, 100, 128–35,
  137
Martin, A., 6
Mateo, N., 112
Malton, P.J., 6, 88, 93
maturity periods, 6, 40, 52–3, 62, 82, 86, 87,
  91, 92, 124, 149
Maurya, D.H., 3, 93
Mbewe, Martin, 13, 99, 100, 143–5
McCorkle, C.M., 6
mechanization, 32, 147
media, 163–4
medicinal plants, 66, 144–5
Mediterranean, 68
Mekbib, Hailu, 9, 61, 62, 78–84
Mellor, J.W., 5
'memory banking', 105, 109
Merrick, L.C., 2, 57, 68, 69
Mesoamerica, 68
Mesopotamia, 68
methodologies, 1, 4, 9, 11, 19, 24, 48, 93–4,
  99, 115–7, 120–1, 157–9 *passim*
Mexico, 3, 4, 175
migration, out, 35, 57, 131
Millar, David, 17, 18, 44–50
millet, 12, 86, 87, 149, 160
monocultures, 2, 65, 85, 146, 148, 162
monopolies, 12, 155, 156, 171, 174, 178
Mooney, P.R., 4, 108, 155, 156, 172–8
Moss, H., 144
Muller, E., 11
multiple enterprises, 32, 34
Mushita, T. Andrew, 61, 62, 85–8, 90
Myrada project, South India, 11, 100,
  129–35
myths, 30, 160–1, 163, 171

Nabham, G.P., 68
Nazarea-Sandoval, V.D., 92
Ndiweni, M., 139
Nepal, 10, 69

Netherlands, 26
networks, farmer, 6, 11, 21, 22, 139
New Zealand, 175, 176
Newbouldia laevis, 40
NGOs, xii, 9–13 *passim*, 25, 26, 70, 76, 99,
  100, 108, 128, 129, 133, 136–42, 150, 172
  see also individual headings
  grassroots service organisations, 136, 139
Norman, D., 5
North America, xii, 65
NRC, 1, 2, 115
nutrient cycling, 37–40 *passim*, 68
nutrition, 155, 160–1, 163
Nye, P.H., 38
Nyerges, A.E., 41, 56

oats, 112
Obilana, T.A., 92
ODI, 137
OECD, 175, 176
officials, 24–5
oil crops, 65, 85
okra, 28, 37
Oldfield, M., 2
Oomkes, F.R., 20
van Oosterhout, Saskia, 61, 62, 89–95
Opole, Monica, 155, 157–64
ORAP, 139
overgrazing, 147
ownership, 158, 173–5 *see also* intellectual
  property
Owusu, D.Y., 42
Oxfam, 10

participation, farmer, 4–11, 18, 42, 43, 62,
  102–4, 128–35 *see also* appraisal; research;
  technology
'passport data', 104
patents/patenting, 4, 12, 13, 26, 156, 166,
  171, 173–5 *passim*, 178
peer pressure, 47–8
pepper, hot, 28, 31, 37, 80
Peru, 6, 44–5, 106–8, 112, 114–18, 174 *see*
  also potatoes
  Seed Production/Distribution Project,
  106–7
pesticides, 2, 7, 22, 32, 64, 85, 147, 148,
  160, 162
pests, 2, 22, 57, 79, 81, 85, 86, 122, 123,
  162
  resistance, 79, 82, 87, 91, 108, 122, 125,
  126
Philippines, xii, 11, 23, 92, 105, 109–10
  UPWARD project, 105, 109–10
planning, 11, 19, 29, 100, 162

203

local, 1, 12, 22, 76, 166–71, 174–8 *passim*
  see also landraces
open-pollinated, 12, 85, 86, 149
wild, 1, 64, 67, 68, 74, 78, 84, 92–3, 106,
  143–5
Vaughan, D.A., 2
Vavilov, N.I., 72
vegetables, 28–30 *passim*, 65, 66, 79, 80,
  109, 112, 137, 139, 155, 178
  Asian-Research and Development Centre,
  178
  Indian Vegetable Project, 158–9, 163–4
veld products, 100, 143–5
veterinary medicine, 137
vocabularies, 22
Voss, J., 3

Warren, D.M., 1
waterlogging, 126
watershed development, 11, 128–35
  Myrada PIDOW project, 11, 100, 129–35
Watts, M., 53
weather, 28, 29, 160
Weber, M., 21
weeding, 6, 32, 33, 37–9 *passim*, 79, 91, 92
weeds, 1, 3, 32, 33, 35, 37–9 *passim*, 42,
  68, 85, 91, 140
weevils, 91, 125
Wehmeyer, A.S., 144
Weija Irrigation Project, 17–18, 27–34
Weiskel, T.C., 68
Wellard, Kate, 1–13, 99–101, 136–42

wetlands, 54, 68, 149
wheat, 2, 79, 80, 83, 171, 175–7
Wilairat, A., 7
Wilde, J.C., 5
Wilke, P.J., 68
Wilkes, H.G., 3, 68
Williams, J.T., 68
women, 10, 18, 30–2, 34, 36, 53, 75, 79–82
  passim, 91, 106, 109, 117, 125, 129, 137,
  141, 150, 155, 158, 159, 163 *see also*
  gender relations
Wood, D., 2, 4
Worede, Melaku, 9, 61, 62, 78–84
World Neighbours, 139
worldviews, farmer, 17, 18, 21–3 *passim*,
  26, 44–50
WRI, 2
yams, 2, 109
ye-eb, 84
yield, 28, 61, 62, 64–6 *passim*, 70, 85–7
  passim, 92, 170
  maximization of, 2, 4, 5, 65, 66, 70
young people, 53
Yurjevic, A., 141
Zimbabwe, 61, 62, 85–95, 139
  communal areas, 62, 85–95
  ENDA project, 62, 85–8
  Seeds Action Network, 12